Quality Control in Toxicology

Quality Control in Toxicology

Edited by
G. E. Paget

Director, Inveresk Research International

Published by

MTP Press Limited
St Leonard's House
St Leonardgate,
Lancaster, Lancs
England

Copyright © 1977 MTP Press Limited

Softcover reprint of the hardcover 1st edition 1977

ISBN-13: 978-94-011-7184-7   e-ISBN-13: 978-94-011-7182-3

DOI: 10.1007/978-94-011-7182-3

Phototypeset by Fyldetype Ltd., Preston
and printed by Blackburn Times Press Ltd., Blackburn

Contents

SECTION THREE

The Independent Expert
Chairman: J. McL Philp

SECTION FOUR

The Research Contractor
Chairman: J. McL Philp

Quality Control in Toxicology *was sponsored and organized
by Inveresk Research International. The meeting was the
second in their series "Topics in Toxicology" aimed at
producing meetings on subjects of topical importance to
toxicologists.*

List of Contributors

SHERWIN GARDNER, BMEng
Commissioner, Food and Drug Administration,
5600 Fishers Lane, Rockville, Maryland 20857, USA

JOHN P. GRIFFIN, PhD, BSc, MB, BS, LRCP, MRCS
Principal Medical Officer, Medicines Division,
Department of Health and Social Security,
Finsbury Square House, 33/37A Finsbury Square,
London, EC2A 1PP
Tel: 01-638 6020 ext 385

Professor JOSEPH J. C. JACOB, MD
Head of the Laboratory of Pharmacology and Toxicology
Institut Pasteur, 25 rue Docteur Roux, Paris 16e, France

Professor PERCY LINDGREN
National Board of Health and Welfare,
Department of Drugs, Box 607, S-751 25, Uppsala, Sweden
Tel: 010 4618 100360

BRIAN J. LEONARD, FRCPath, MB, ChB, MD
Manager, Safety of Medicines Department, Imperial Chemical
Industries Limited, Pharmaceuticals Division,
Mereside, Alderley Park, Macclesfield, Cheshire
Tel: 0996 6 2828, Telex 669095/669388

WILLIAM M. MERINO, PhD, BS
Director, Domestic Pharmaceuticals,
Regulatory Affairs Department, Searle Laboratories,
Box 5110, Chicago, Illinois 60680, USA
Tel: 010 1 312 982 7000, Telex: 72/4437 or 724430

PETER R. B. NOEL, MB, BS
Medical and Scientific Director, Huntingdon Research Centre,
Huntingdon, PE18 6ES
Tel: 0480 890431, Telex 32100

G. EDWARD PAGET, MD, DCH, FRSE
Director, Inveresk Research International, Inveresk Village,
Edinburgh EH21 7UB, Scotland
Tel: 031-665 6881 Telex: 727228

Professor DENNIS V. W. PARKE, PhD, DSc, FRIC, FIBiol, FRCPath
Department of Biochemistry, University of Surrey,
Guildford, Surrey, GU2 5XH
Tel: 0483 71281, Telex 85331

JACK McLEAN PHILP, MRCVS, FIBiol
Environmental Safety Officer, Research Division,
Unilever Limited, PO Box 68, Unilever House,
London EC4P 4BQ
Tel: 01-353 7474, Telex 28395

HARALD REINERT, MD
Chief Toxicologist, Inveresk Research International,
Inveresk Village, Edinburgh EH21 7UB, Scotland
Tel: 031-665 6881 Telex: 727228

Professor Sir ERIC FRANK SCOWEN, MD, DSc, FRCP, FRCS, FRCPE, FRCPath
44 Lincoln's Inn Fields, London, WC2A 3PX
Tel: 01-405 5580

DONALD EDWARD STEVENSON, BSc, BVSc, MA, PhD, MRCVS
Director, Shell Toxicology Laboratory (Tunstall),
Shell Research Limited, Sittingbourne Research Centre,
Sittingbourne, Kent, ME9 8AG
Tel: 0795 2444, Telex: 96181

Sir GORDON WOLSTENHOLME, OBE, MA, MB, BChir, MRCS, FRCP, Hon FACP, FIBiol
President, The Royal Society of Medicine,
1 Wimpole Street, London, W1M 8AE
Tel: 01-580 2070
NB Sir Gordon works at The Ciba Foundation

Series Preface
Topics in Toxicology

Toxicology is a science that stands at the intersection of several interests
and disciplines. These intersecting forces are by no means all scientific
since some are legal and some are commercial. All have valid things to
say about the conduct and interpretation of toxicity experiments. The
practising toxicologist must bear all these sometimes conflicting forces
in mind as he carries out his duties. This is especially true, of course,
of the toxicologist in industry.

Toxicology is also a field in which contract research particularly
flourishes and a number of major contract research companies have
established over the years a reputation for contributing usefully to the
practice of this skill. These contract research organisations are particularly favoured to develop an appreciation of the problems of the industrial toxicologist, since it is very common for such a company to
service the toxicological needs of companies in several sectors of industry
producing new chemical compounds, and the contract research company,
therefore, is aware of a wider diversiy of problems than probably affects
a single toxicologist in one industry.

Among the problems of which all toxicologists in industry are aware
and particularly so those in contract research companies, is the relative
paucity of fora for adequate toxicological discussion. There are, it is
true, at least two major societies concerned with problems of toxicology,
the European Society of Toxicology and the Society of Toxicology in
America. Although performing a useful function, these societies are now
so large, and cater for such a wide spread of interests and contain so
many individuals whose interests in toxicology is at most peripheral,
that many toxicologists feel they do not provide an adequate platform
for discussion of the most immediate and important issues affecting
toxicology at any one time. The very size and complexity of these
societies means that the organization of their meetings and the topics
for their symposia must be chosen many months, and indeed years, in
advance. Thus issues of current importance cannot usually be dealt with
while they are still live.

Inveresk Research International, a contract research company operating, among other areas of contract research, in the general field of
toxicology, feels that there is a need for small, highly specialized and
specific, meetings, arranged at relatively short notice to cover important

current issues in toxicology and has instituted the series 'Topics in toxicology' to meet this need. The subjects for these meetings will, as far as possible, be ones of immediate concern to practising toxicologists in industry. They will bring together, as far as possible, the most authoritative opinions bearing on the particular subject. Subjects will be chosen within the general constraint of current interest, to be ones which stand at the intersection of at least two of the several strands of interest that control industrial toxicology.

The first meeting in this series dealt with toxic hazards in industry seen in the light of the Health and Safety at Work Act. This was of particular interest to individuals in the chemical industry who have to cope with this aspect of safety now focussed by the Health and Safety at Work Act.

The second meeting was equally topical and concerned Quality Control in Toxicology in the light of the standards of Good Laboratory Practice published by the FDA only two months before the date of the meeting.

It is IRI's intention to mount similar meetings in the series as suitable subjects for them become interesting. There will probably be at least one per annum and possibly two. Anyone wishing to be made aware of either the meeting or the subsequent publication should write to 'Topics in Toxicology', Inveresk Research International, Edinburgh, EH21 7UB, Scotland.

G. E. Paget

Foreword

When we use toxicology to mean the safety evaluation of new chemicals of all sorts we describe a science that, in addition to the peculiarities described in the Series Preface, has a number of other characteristics. Perhaps the most striking of these, and one in which it differs from almost any other scientific endeavour, is the fact that it deals with very large volumes of data. A relatively small experiment in safety evaluation may readily generate in excess of quarter of a million separate items of data. Although, obviously, the relative importance of each item varies, nevertheless each item does have some weight in the final outcome of the experiment, either directly as an individual observation or in summation as one statistic to be considered with large numbers of others of like kind. Each item eventually comes to be represented in a decision of which the most characteristic form is to use a compound or not to use it. In this sense, therefore, it is true that the decisions that are made about the compounds to be introduced to the human environment depend, albeit perhaps in some cases to a very small degree, upon the accuracy of each item of information that is represented in the final decision matrix. Those that must consider the validity of the data, and that themselves must decide or advise upon the decision to be made upon the basis of that data, have come to place great reliance upon the accuracy of each single item and it is not a very large step from this position to believe that each single item has exactly the same weight in the final decision process as any other. It is therefore particularly disconcerting to this attitude of mind when it can be shown that certain items of the data relating to a particular experiment are inaccurate, irreproducible or indeed, in some cases, dishonestly falsified. Unfortunately, it is also true that the gathering of such very large amounts of data must, to a very large extent, be placed in the hands of technicians and routine workers rather than in the hands of highly motivated scientists. Thus, while it is certainly true that such technicians may not be motivated to falsify the data in quite the flamboyant way of certain notorious scientific scandals of recent years, it is also true that some junior technicians are not motivated to handle each individual, and apparently unimportant, measurement with the same care that a more highly trained scientist might do. It is also true that the generation of of these very large amounts of information must inevitably carry, as it were, a risk of replication errors, the mere copying of 250 000 separate items of information from a primary record to a secondary record will, inevitably, whatever mechanism of copying is adopted, lead to transcription errors.

No one who has been acquainted with the field of toxicology for many years was surprised when it turned out that, under intensive investigation, a number of toxicological experiments were shown to have serious errors in the accessibility and reproducibility and validity of the data. Indeed, it would be surprising to such an experienced toxicologist if, under intensive audit, errors were not found in such a large data collection. However, coming at a time when there is a marked tendency for the consumer to distrust the scientist and when the whole structure of consumer protection from toxic chemicals is under public scrutiny and when, moreover, the science of toxicology has been made to look either inefficient or foolish by failures to predict catastrophes and by apparently absurd inconsistencies, these valid criticisms of the audit of data in toxicology provoked a disproportionately large reaction. It became clear, partly as a result of regulatory action in the United States by the FDA, and also by the general feeling throughout the world, that inquisition into the way in which toxicology data are gathered would be proper, and that controls should be placed upon the way in which toxicology laboratories carried out their experiments. Even the most individualistic of toxicologists would agree that there can be little complaint about the ostensible purpose of good laboratory practice, which is to ensure that experiments are carried out in the way called for by the protocol defining them and that the data which it is claimed in the protocol would be collected are, in fact, collected in an accurate and reproducible form.

The areas of debate spelt out in the symposium on this subject reported in this volume define some of the problems working scientists expect to arise.

It will be clear that this is but the opening of a phase of toxicology of which it is difficult to foresee the end. It is certain, of course, that the preliminary steps being taken in the United States will be followed by similar actions or perhaps reactions in other countries and that the practice of the toxicology of environmental chemicals will be profoundly altered. It may well be that in some years' time a similar symposium will come to wonder whether the impact of good laboratory practice has been wholly beneficial. At this stage it is too early to make such a prediction although it is certain that many practising toxicologists have misgivings, whereas others feel that the misgivings may be justified but will be more than counterbalanced by benefits other than the relative trivia that are the subject of regulation by GLP that may arise. One of the most controversial matters raised by the proposed GLP is the requirement that laboratories submitting data to the FDA should establish an internal quality assurance department analogous to that which controls manufacture of many products. This is an innovation in most toxicological laboratories, and has given rise to much controversy already. IRI has, itself, in the last few months appointed a Quality Assurance Manager, whose duties will be to ensure that the internal regulations laid down for the conduct of our studies are complied with by our own staff. His department will be also the major source of data audit and the controllers of archival records of toxicology. We believe that the institution of such a department will do much to ensure that data are collected and retained in readily organized and recognized form and that his department will free the scientist actually conducting the experiment from relatively trivial concerns, so that he can devote his attention to more fundamental issues.

There has been considerable debate about the requisite qualifications for Quality Assurance Managers and we have taken the step of placing in charge of IRI's Quality Assurance Department neither a toxicologist nor a non-scientist but rather a trained scientist whose speciality has been quality assurance in a highly technical environment. This seems to us to be the correct decision as between the various options and it is to be recognized that quality assurance and methods of achieving it is becoming a highly organized and important speciality in its own right. The Quality Assurance Manager will not, of course, have the right of querying either the interpretation of scientists nor those aspects of the operation that are strictly within the purview of the qualified scientist. However, he will be drawn integrally into the design of protocols and even more into the design of recording procedures in order to ensure that such matters are handled in such a way as to enable quality assurance to be properly applied.

Whatever toxicologists may think there is no doubt that our discipline has entered the era of good laboratory practice. What its impact on the safety evaluation of chemicals may be remains to be seen.

G. E. Paget

SECTION ONE

The Regulatory Viewpoint
Chairman:
Sir Gordon Wolstenholme

1

Maintaining the creative balance
Sherwin Gardner

This gathering of experts is a clear recognition of our common interest in, and concern with, assuring the safety of chemicals that have a useful place in society. Just as science cannot be contained by political boundaries, and just as commodities flow between nations, so too do the problems of assuring the quality of those commodities and the conduct of science. Certainly, the US Food and Drug Administration is no stranger to international collaboration to assure safety and, more importantly, to co-operate in efforts to raise the standards by which safety is measured. For example, we have formal reciprocal drug plant inspection with both Sweden and Canada, and informal arrangements with other nations. We have agreements with The Netherlands, France, Belgium, and New Zealand concerning the sanitary production and laboratory analysis of dry milk; and with Canada, Japan, and Korea on sanitary control of shellfish production. These arrangements are, in my view, unequivocal evidence of the genuine nature of the international commitment to safety. Today, as one more reflection of that commitment, we meet to explore the subject of quality control of toxicology. I congratulate those who conceived and organized this symposium, for there could not be a more timely, or central, subject than that we have been asked to address.

Toxicology provides society with the basic mechanism to evaluate the risks and benefits of chemicals introduced purposely or by accident into a biological system. The quality of the methods chosen, and the existence of adequate control over those methods, are as vital to toxicology as is the accuracy of the yardstick to civil engineering.

The Food and Drug Administration is charged by the American people with serving as their primary agent for safeguarding the food and drug supply. Thus, the FDA is particularly sensitive to the need for quality assurance in toxicological testing of chemicals as well as in the manufacturing of chemicals. Indeed, with the passage of time and the evolution of the Agency, that sensitivity has greatly increased. Throughout much of its institutional life, the Food and Drug Administration carried out legislative mandates that, by and large, limited the Agency to 'catch-the-offender' actions. A product was shipped and sold, and, if FDA discovered a violation of law, it moved in court against the product and its maker. The burden of proving that a violation of law did, in fact, occur was squarely on the shoulders of Government. As time passed, the inadequacies of this approach became increasingly obvious. As a result, the philosophy of regulation, as reflected in amendments to our basic law, changed from corrective after-the-fact action to preventive before-the-fact action, i.e. to product evaluation and approval before marketing could take place.

Not only was regulation converted to a preventive strategy, it was changed also in terms of responsibility and burden of proof. Formerly, the FDA's position usually was that of a petitioner. It was the Agency's responsibility to initiate scientific and legal proceedings to review the safety of products subject to its jurisdiction, but only after marketing. The burden of proof was largely on the Food and Drug Administration. Now, the manufacturer of a product has the responsibility to initiate a safety review and submit evidence of safety prior to marketing in most cases. Thus, the burden of proof was shifted from Government to the manufacturer.

In addition, the law also introduced standards for safety and effectiveness that are grounded in scientific process. These standards are admittedly broad, but that is appropriate considering that we will have to accommodate to change brought about by scientific discovery. For example, for a food substance to be generally recognized as safe among experts qualified by scientific training and experience to evaluate safety, it must have been shown through scientific procedures to be safe under conditions of intended use. For new drugs, the safety and effectiveness determinations also depend upon experts qualified by scientific training and experience. Safety and effectiveness claims for drugs must be supported by evidence consisting of adequate and well controlled investigations conducted by scientifically trained and experienced persons. In short, science has become the foundation of the regulatory structure, and particularly the science of toxicology.

I would add that an important component of that foundation is the professionalism of those who gather, record and submit scientific data. The policy of the FDA necessarily has been that unless there is a compelling reason to believe otherwise, we would proceed from the assumption that the foundation was intact, and that the evidence submitted to support an application reflected professionalism and science of the highest order. Based on this assumption, our practice has been to examine the results of scientific studies, as well as the adequacy of the procedures and methods described in written reports of those studies. With rare exceptions where cause was demonstrated, we did not examine work in progress. To a major degree, this simply reflected the realities of resource limitations: To monitor on-going laboratory and clinical research on a routine basis would require manpower and funds in what was judged to be a non-cost/effective manner. Certainly, to reproduce everyone's research would be even less cost/effective and an intolerable drain on available funds and scientific manpower.

The approach seemed to fit the concept of scientific integrity as well as our American concept of a proper and enlightened balance between productive free enterprise and Government regulation. However, recent events have caused us to depart from this policy of looking primarily at results and protocols and not at work in progress. What we have found in several laboratories produced what conservatively can be described as extreme concern.

For example, we have found experiments that were poorly conceived, carelessly executed, inaccurately analysed, and poorly recorded. To be more specific:

(1) We have been unable to find original autopsy records for test animals.

(2) We have found necropsy data apparently transcribed to new records several years after the event.

(3) We have discovered pathology reports in our files that were inconsistent with the originals.

(4) We have been startled to find instances where experimental animals have expired, only later to begin life anew.

These sorts of findings generally bear on the accuracy and adequacy of data recording and maintenance, and would have been of sufficient concern to generate a reaction by FDA to the situation. There were, however, other inadequacies of a fundamental nature that led us to question the entire approach to research by sponsoring institutions. We have found instances where tissue examinations were conducted by more than one pathologist, each of whom came to different conclusions, with only those examinations favouring the chemical submitted to us. We also found that:

(1) Technical personnel were unaware of the importance of adhering to protocol, and of accurately administering test substances and recording data.

(2) Studies were compromised by protocol designs that excluded certain available data. Selected examinations were made of test animals so that comparison of different levels of exposure to the test substance could not be made.

(3) Of particular concern was the apparently casual relationship between sponsors of investigations and contract laboratories who performed some or all of the studies. Sponsors failed to monitor the studies adequately; required pathology examinations were not accomplished, and, in some instances, tissue samples were not collected.

(4) Firms failed to verify in a systematic manner the accuracy and completeness of scientific data in reports of non-clinical laboratory studies before submission to FDA.

In short, we found that we were relying on quality control mechanisms that, themselves, lacked satisfactory controls and quality.

I should like to emphasize that the deficiencies I have mentioned are not isolated instances occurring in a single laboratory or firm. We discovered deficiencies in several laboratories, independently managed, and of the contract research type, as well as those operated by drug manufacturers. While the discoveries during the past year or so led directly to our bioresearch monitoring program, we have uncovered research deficiencies as far back as 1962. It appeared that we were faced with a condition that was widespread and which, if not corrected, had serious implications. Under our laws regulating foods and drugs, invalid data could result in the suspension of marketing for many products and prolong the evaluation of those not yet approved for marketing. You can readily appreciate the resource consequences to private firms and to the Government as well. Thus, this posed a sharp and immediate challenge to our entire system of regulation. Since the adequacy of the scientific foundation was in question, the confidence we should have in our food and drug supply was compromised, and would remain so until the questions had been adequately resolved. I should add that what we had uncovered in terms of weakness in toxicology quality assurance was not confined to the evaluation of substances under FDA jurisdiction. There was a general problem, shared by all Government agencies responsible for regulating or evaluating chemicals, and this was supported by statements from representatives of our Environmental Protection Agency and the National Cancer Institute.

Reflecting the serious and potentially widespread nature of these disclosures and their non-partisan nature, Congress and the Administration joined in providing resources to FDA to establish a control programme. The purposes of the programme are to measure the full extent and degree of the problem and to initiate whatever measures are necessary to restore or establish quality assurance systems in pre-clinical testing laboratories.

Our Bioresearch Monitoring Program, which is what we have named the pre-clinical laboratory surveillance activities, is now under way. It is of broad scope and includes considerably more than surveillance of toxicology testing. Approximately 40% of the resources recently given to bioresearch monitoring is allocated to inspecting laboratories. The remaining resources are directed to monitoring the very important clinical investigation activities that are involved in new drug research, and to conducting a safety review of food additives. One of the first steps we took in executing the programme was to prepare and issue proposed standards for non-clinical laboratory studies. These proposed standards eventually will become enforceable regulations administered by the Food and Drug Administration.

The standards apply to a broad range of laboratory features—personnel, facilities, records, operating systems, and other important fundamentals of laboratory operations. I believe that many people experienced in laboratory operations, or other scientific and technical enterprises, will read these standards and conclude that much of what we have stated is obvious and is generally taken for granted. For example, on the subject of organization and personnel, we require employment of suitable numbers of persons with appropriate education, training and experience to perform the activities of the laboratory. We further require that there be a study director responsible for each study, and that this director be a qualified scientist. The director would be responsible for the overall conduct of the study, and would assure that the approved protocol is faithfully followed. Lest there be any doubt about the specific aspects of the director's responsibility, these are enumerated in the standard. The purpose of this requirement is to assure accountability for the control and validity of the study.

One of the key features of the laboratory practices standard involves establishment of a quality assurance unit. Experience has shown that detailed protocols and written standard operating procedures alone will not ensure the quality and integrity of the results of a non-clinical laboratory study. The word may be the beginning, but it does not necessarily and automatically translate into the deed. A positive mechanism is needed to assure the quality and integrity of results obtained from laboratory studies. In this regard, the quality assurance unit complements the functions of the study director by assuring that facilities, equipment, methods, and records conform with applicable standards. It is important to note that the primary role of the quality assurance unit is that of monitoring the performance of the research. The regulations do not place, and are not intended to place, with the quality assurance unit the responsibility for accepting or rejecting a specific study design or its results, or for approving or rejecting standard operating procedures.

This reflects our philosophy about the entire bioresearch monitoring programme. We are not going into the business of taking on the responsibility for assuring the quality of research. That still remains where it should be—with the laboratories and manufacturers. What we *are* doing is making sure that those who conduct such research have a mechanism for assuring themselves and us of the rigorous application of study protocols and the accuracy of data collection analysis, and reporting.

I suspect the requirement for a quality assurance unit may be one of the more controversial aspects of this proposed standard. Perhaps, it would be appropriate to share with you some of my own experience in this area. While with an aerospace engineering firm, one of my responsibilities was for a laboratory quality assurance function similar to the one described in our proposed standard. Having been educated and trained as an engineer, my laboratory experience relates primarily to working with other engineers, physicists and mathematicians. My observations of biological scientists, however, lead me to conclude they behave similarly in a research laboratory environment; that is, they get enthusiastic about and caught up in their work. Sometimes, this causes them to omit the rigorous attention to detail and procedure that valid scientific study demands. The most frequent casualty is record keeping, which becomes sloppy and inaccurate. Indeed, I have observed data being recorded on shopping bags, match books and in one case on a piece of lumber! (I think that fellow took instructions about keeping a log book too literally.) Other common deficiencies in laboratory practice are the use of uncalibrated instruments, and failure to use good measurement practices. My point is that there is a role for a laboratory quality assurance unit; that quality assurance in the laboratory is a do-able job and one which needs doing.

My remark about record keeping calls attention to another of the important provisions of the proposed standard—we have placed a great deal of emphasis on records and reports. These are the products, so-to-speak, of test laboratories, and provide the basis for our evaluation and approval of drugs and food additives. Complete and accurate records also are essential if it becomes necessary to re-evaluate some aspect of the substance because of new scientific information. Obtaining reliable research records, thus, should be a mutual objective of manufacturers as well as regulatory agencies. Both wish to obtain and continue public credibility in their respective institutions and both wish to be efficient in their use of resources.

I will not further describe the details of the laboratory practices standard. I believe they are well set out and discussed in the proposal. It would be useful, however, to discuss the status of our proposal and the implications of the standard. I have been referring to the standard as a proposal, and also stated that it would become a regulation to be administered by FDA. For those of you who may not be familiar with administrative law practices in America, regulatory agencies generally must establish regulations in a two-step rule making process. The first step is to publish a regulation and obtain public comment on it. The second step is to publish a final regulation after having considered the comments offered and making appropriate changes. Sometimes, the comment is obtained entirely by written response to the Agency; sometimes, hearings are held to establish a record which the Agency will use to make a final determination. In this case, we have taken only the first step so far. We have asked for written comment and, in addition, will hold a public hearing next month. The hearing will afford the opportunity for an exchange of views on the scientific soundness and practicality of the proposed regulations. All comments, whether written or obtained at the hearing, will be taken into consideration in moving from proposed to final regulations. Because of the import of, and extensive interest in, this subject, I expect the comments will be extensive. We hope to issue a final regulation by next July.

There are two encouraging signs I would like to take note of. First, we received acknowledgement from the drug industry that a problem does exist. Independent proposals for guidelines describing good laboratory practices were submitted by the Pharmaceutical Manufacturers Association and by the G. D. Searle Co., in advance of our proposal. We are looking forward to receiving informed and professional viewpoints from many other interested parties on the substance of our proposal. Second, for about 2 months, we have been conducting inspections of a randomly selected group of non-clinical laboratories. These inspections are intended to gain additional information about current laboratory practices and also to evaluate the application of the proposed laboratory practice standard. While it is too early to reach any conclusions, preliminary reports of these inspections are somewhat reassuring that we do not have a disaster of major proportions waiting to be uncovered. A detailed and searching evaluation of these selected inspections will, of course, be completed before we issue final regulations and initiate inspections on a larger scale. Regardless of the outcome of the selected inspections we will conduct intense monitoring of laboratories for about 2 years. After that time, we will be in a position to know precisely the nature and dimensions of the research laboratory problem and what measures might be appropriate for the future. I suspect that many persons in industry and the professions have questions about how the monitoring programme will work. Specifically, who will be affected, and how?

Since some laboratory work supporting applications to market products in the United States has also been carried out by laboratories abroad, and since it is not the intent of the law or our regulations to drive research overseas, the laboratory practice standard will apply equally to all work, whether done in foreign or domestic laboratories. Of course, no laboratory not involved in some way in supporting an application to market in the United States need pay any attention to the proposals and the inspection mechanism. For such laboratories their meaning is academic. For others, their meaning is direct and very real. The meaning also is direct and real to manufacturers.

With regard to laboratories, our inspectors will advise management of any deficiencies uncovered. The strength of the advice and requests for corrective action will depend on the nature of the deficiency. In some instances, we may determine that a specific laboratory study will be unacceptable support for approving an application for marketing a product. In extreme cases of laboratory deficiency, which we hope will be rare, we may disqualify a laboratory as a source of data in support of any application for marketing approval. Obviously, disqualification of specific studies or laboratories has profound implications for both laboratories and manufacturers.

I do not wish to leave the impression that we have pre-judged the case. I believe we will find that many laboratories in the United States and abroad already meet or exceed the Good Laboratory Practices standard. I also believe, however, that some change is needed: The evidence in hand is a clear demonstration of this. Perhaps, some will argue that our proposed approach is too extreme, that other measures less demanding will suffice to bring about the necessary improvement in quality of performance. There are others, however, who advocate even stronger measures. There are, for example, proposals that Government step in and, itself, conduct this research; other proposals would require a third party (someone other than the manufacturer) to do the job. While these proposals are sincere and idealistic, they neglect one basic element. What we are rightfully concerned with is assuring that quality research advances scientific creativity.

It is obvious that no one has a monopoly on quality research. All government research is not necessarily superlative, nor is industrial research of poor quality merely because it is profit-motivated. The key is not who does the research, or where, or what name is over the entrance to the building. The key is quality supporting creativity. Quality is the attribute common to all research, whether publicly or privately carried out. I believe that the approach we in FDA are taking will achieve quality in research while maintaining a system that has a proven track record of creativity.

Society has a major stake in assuring the continued health of its creative systems. One way to assure that health when the need arises is through prompt application of appropriate therapeutic measures, such as those I have described. Such prompt application is the best way I know to avoid, perhaps, extensive and otherwise needless surgery. I urge all who are involved in the research that supports safety decisions about chemicals to take a close look at the way such research is being conducted. Examine whether there are proper mechanisms in place to assure the quality we have taken for granted and now must take steps to guarantee.

2

A Swedish regulatory viewpoint
Percy Lindgren

The title 'Quality control in toxicology' may convey a different meaning to different readers. The topic has been discussed at several meetings recently and in different connexions, and we know that some toxicology investigations have been heavily criticized for not having been performed according to 'good laboratory practice'. It thus seems well worth while to discuss the topic at some length. This is therefore the right conference at the right time, and may be at the right place.

There is certainly a great deal that can be said about the topic from the point of view of a drug control authority. We have just listened to a description of the detailed inspection programme that the FDA has worked out in order to make sure that experimental toxicological results will have a certain degree of reliability. I do not intend to discuss the various aspects of this programme in detail or its implications, but, as will be seen below, I do share several of the views expressed. I will therefore try to look upon the topic 'quality control' from a somewhat different angle and to broaden the definition a little. My point of view is, of course, coloured both by my scientific experience and by my experience in reviewing other scientist's results and having had to make up my mind in a number of specific cases from the drug control point of view. Although some overlapping cannot be avoided, several levels of

'Quality control in toxicology' can be distinguished. I have chosen to comment upon the following ones:

1. THE QUALITY OF A SINGLE OBSERVATION

This means the accuracy of the measurements and is primarily a question of equipment and technique, and in some respects, of carefulness of the technician. This aspect is quite obvious to everybody. If for example we measure the electrolyte concentrations in urine, we use carefully calibrated instruments; if we determine the amount of a radioactive drug metabolite in a blood sample, we check the background radioactivity; if we produce a histological slide of a tumour tissue, we make sure that it is properly prepared, mounted and stained, so that no unrecognized artefacts will appear.

2. THE QUALITY OF THE RAW DATA, INCLUDING STATISTICAL EVALUATION

Such factors as uniformity of experimental conditions, reproducibility of the results are important. Several aspects come to mind, particularly for long term toxicity tests. Adequate specifications should exist for the substance (batch) being tested. The specifications should define the substance in terms of its purity (limits for impurities), solubility, particle size and any other aspects which may bear upon its biological effect. The stability of the substance in the vehicle used for its administration in the toxicological tests should be documented. If the specifications are altered, for example by alteration of the manufacturing process, the relevance of those tests performed on earlier manufacturing batches must be judged with reference thereto. Such questions as how the dosage is expressed and plasma level correlated, if the drug is mixed in the food; what precautions are made to assure that the drug is not changed during storage of the food; how much is the daily intake by each animal, if several animals are housed in the same cage, and so on, must be answered. It is also important not to use statistical methods and draw conclusions beyond what the biological characteristics of the data permit. I think these two levels of 'quality control'—the single observation and compilation of raw data—have been so extensively covered in the first communication by Mr Gardner, that there is no need for further discussion.

3. THE BIOLOGICAL QUALITY OF THE DATA

Do we prefer genetic homogenieity or heterogenieity? Do we use special strains of mice and rats for different types of tests, or do we use the same strain for which the laboratory has acquired a basic knowledge? Most experiments in mice and rats are carried out with albinos so a possible melanin affinity will not be detected, and the data collected will lack this particular quality.

It is probable that an effect is more likely to be detected if pure bred beagle dogs are used instead of mongrel dogs, and thus that a lesser number of animals and smaller costs may be needed. But will the data be more reliable generally speaking, and thus have a higher quality? It is not possible to say that toxicity data obtained in beagles have a higher quality than those obtained in mongrel dogs, but they do have a different quality. Under all circumstances it is important to define the genetic specifications. The biological quality is in many respects a question of knowing and specifying the genetic limitations. The importance of knowing and specifying the limitations also holds true for other factors, for example the environment. In this context environment is generally defined as 'everything that is not genetic'. Many examples showing the significance of environmental factors could easily be given. Two classic ones well known from the literature may be enough. The first example, (Figure 2.1) shows the influence of room temperature on the LD_{50} value in mice given chlorpromazine. Even within such a narrow range as 20–30°C the LD_{50} value changes almost ten-fold, from low levels at both the beginning and the end of the interval to a peak at 28°C. This means that at certain room temperatures even small variations, may be 1/2 or 1°C, can considerably influence the results of the experiment. The other example Table 2.1 of environmental factors in toxicity

TABLE 2.1   Tumour incidence in mice. Effect of different dietary conditions

Group	Number of mice	Number per cage	Weight of diet per day	Survival time	Number of tumours
1	40	1	4g	18 months	4
2	40	1	5g	18 months	4
3	40	1	Ad lib.	18 months	32
4	40	5	Ad lib.	18 months	23

No 'test substance' given to any group. Swiss albino males. Standard pelleted diet. Carcinogens possibly present in trace amounts: 3,4-Benzopyrene, Dimethylnitrosamine N.B. Aflatoxin not present in detectable amounts. (Roe and Tucker (1974). *Excerpt. Med. Int. Congr. Ser.* **311**; 171)

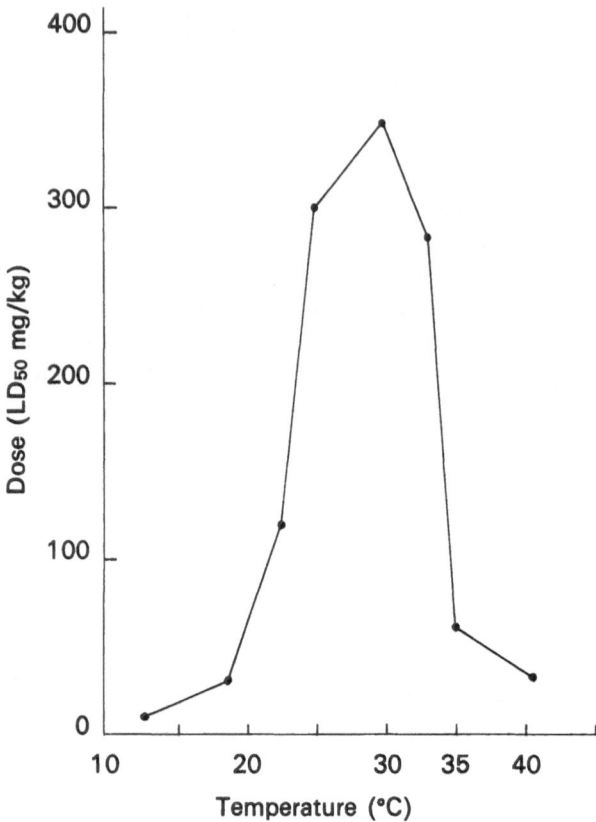

Figure 2.1   Effect of room temperature: LD_{50} variation in mice given chlorpromazine Friis C. W. (1966). Forsøgsdyrkunskap, Jydsk Teknologisk Institut, Aathus

tests is taken from the other extreme in terms of duration, from a test for carcinogenicity. It shows a remarkable increase in tumours if the animals are given as much food as they like, compared with groups given a restricted amount of the same standard diet. The authors attribute the results to the possible presence of carcinogens in trace amounts in the food. Other environmental factors, sometimes neglected, are the use of insecticide in the animal quarters, or the administration of antibiotics. Apart from the fact that these agents may have a toxic effect on their own they can influence the animal's reaction to the tested substance, e.g. by means of enzyme induction.

4. QUALITY IN PLANNING, PERFORMING, REPORTING, EVALUATING

This involves anything from the general design of the experiment, quite essential for the value of the toxicological data and conclusions to be produced, to the strict laboratory discipline to assure that the individual protocol is correctly followed, this also being quite essential to the quality of the experiment. Important factors are properly trained and experienced personnel, as well as good working conditions. Regulatory authorities often like to set up schematic recommendations and guidelines for all types of toxicity test, and scientists undertaking the work sometimes look upon those as a check list. Each drug has its own characteristics, however, and the toxicological profile has to be established by an individual design of the total toxicology to be carried out, rather than by following a conventional check list. Sometimes a decision not to perform a certain routine test may make it possible to allocate resources to a more relevant type of experiment and thus improve the quality of the latter. Modern techniques for studying the pharmacology of a new drug in great detail, particularly the mode of action at a cellular level, may often facilitate the rational design of toxicity tests. Once the toxicity test has been designed and the protocol has been drawn up it is essential to perform the experiment strictly according to the plan. As regards laboratory conditions, laboratory discipline and strict adherence to the protocol, no guidelines can be more detailed than the FDA manual 'Good Laboratory Practice'.

From the viewpoint of a regulatory authority there is often reason to complain about the quality of reporting and evaluating. Sometimes the reports consist of raw data as collected without any attempt at interpretation or conclusions. The following should always be specified and discussed: animal strain, method of administration and dosages; solvents or suspending agents; the animals by number, sex, age and weight; negative controls and where applicable reference group and positive controls. If a test, generally considered as a routine test for a new substance, has not been performed, reasons should be given why it has not been done. The quality of a long term toxicity experiment does not only depend on the final result; observations carried out during the course of testing are often as important as the histopathology at the end. As a pharmacologist I wonder why more functional (pathophysiological) tests are not undertaken during long term studies? Consider the following example: How do rats in a long term test on adrenergic beta-blockers respond to physical exercise? To what degree can their heart and circulatory system respond in acute situations and in a daily training programme? I cannot remember any new beta-blocker that has adequate information from animal experiments covering monthly functional changes of that kind during a two year exposure to the drug for example.

5. QUALITY—PREDICTIVE VALUE OF TOXICOLOGICAL DATA

Toxicology experiments in animals are generally not performed for their results *per se*. They are solely concerned with the degree of prediction they permit for man. It seems therefore reasonable to talk about a quality aspect of the predictability. The choice of animal species in toxicology is generally acknowledged to be difficult. It is still quite rare for adequate exploratory pharmacokinetic studies to have been reported in a number of animal species (and man) prior to the decision on long term toxicity studies in the documentation filed for the registration of a new drug. Yet it is well known from the literature that great differencies do exist between different species in drug metabolism. A classic example is illustrated in Table 2.2.

Other factors which may improve the predictive quality of toxicology data are a number of quantitative parameters: number of animals, number of dosages, number of species, duration of a long term test. These factors have in the past been a matter of much discussion, not least because the view of them held by drug authorities and other decision-makers may mean a great deal for the costs of the toxicological programme of a new drug. I doubt that anybody will question the conclusion that any increase of these parameters is likely to increase the predictive quality and thus facilitate the safety evaluation. But at the same time we

TABLE 2.2   Plasma concentration of ICI-33828 (Methallibure) (to produce a biological effect on the pituitary) in all the species: 3 μg/ml

Species	Dose (mg/kg)
Rat	20
Rabbit	200
Monkey	15–20
Dog	10
Sheep	10–15
Man	1

Duncan, W. A. M. (1963). *Excerpt. Med.* (I.C.S.) **73**; 67

must admit that the magnitude of the increase of the predictive quality cannot be calculated, and a cost-benefit analysis cannot easily be made.

Another point where the views sometimes held by regulatory authorities has been criticized is the tendency to ask for extended toxicity studies. Some toxicologists have argued that no toxicological effects will show after the initial 3 months provided adequate dosage levels are chosen. The present Swedish regulations require at least a 6-month chronic toxicity tests for drugs intended for long term use (more than one month). Table 2.3 is a summary of two separate studies of the chronic toxicity of tranexamic acid in rats. No symptoms except diarrhoea was seen in the 4-month study at any dosage, whereas hyperplasia of the bile ducts appeared in the 22-month study.

Another aspect of the predictive quality and hence the safety evaluation is the possibility for the reviewer to place the new drug in relation to allied previously known drugs. Unfortunately the use of a reference substance is often neglected in toxicity studies. The term 'reference group' refers to animals given one or several dosage levels of a reference substance, i.e. one of the better-known drugs in the same chemical or therapeutic category. It should not be confused with the term 'positive control group,' i.e. animals given a substance that has a well known effect on the organism, and serves as a check that the animals have the ability to respond during the present experimental conditions.

It is realistic to admit that, no matter how high the pressure for quality in toxicity experiments is, the prediction of toxic effects can never be perfect. It is therefore necessary to have a system where the risks in man are minimized when a new drug is introduced. A way to

TABLE 2.3   **The effect of tranexamic acid in rats**

Dose (mg/kg)	Time	Symptom or Injury
0	4 months	0
1000	4 months	Diarrhoea
2500	4 months	Diarrhoea
5000	4 months	Diarrhoea
0	22 months	0
160–195	22 months	0
650–790	22 months	Hyperplasia of the bile ducts
2860–3380	22 months	Hyperplasia of the bile ducts plus neoplastic changes of the intrahepatic bile ducts

In dogs the injuries were not as pronounced. Only raised transaminase values were detected in the highest dose-group, 1600 mg/kg. Ronden **15** (1972), 233 (In Swedish)

obtain this would be, in addition to experience from ordinary clinical trials, to recommend monitored release of new drugs more often than is done today and to work out more sophisticated systems for observing and reporting adverse reactions.

6. QUALITY OF REVIEWING AND POLICY-MAKING

Reviewing the toxicological documentation for a new drug requires generally some sort of communication between the reviewer and the scientists who have produced the various reports. It is important that the reviewer has an adequate scientific training so that the communication can take place on the same level. Under such circumstances the feed-back information from the reviewer, who presumably has experience from a great variety of reports in toxicological matters, to the toxicologists who have conducted the work may have some effect on the quality of the toxicology in the long run. It is also my personal view that the safety assessment by the reviewer should not unnecessarily hamper the development of the new drugs. It should be borne in mind that a scientifically less experienced reviewer may be more bound to say 'no', if toxicological results are complicated and difficult to interpret—a 'no' decision makes him probably feel less uneasy. Making a 'yes' decision is always difficult and potentially dangerous.

In summary I believe that toxicology experiments with a new, biological active substance generally mean looking for the unexpected. It is desirable that the quality of the toxicological data and conclusions are so well controlled that when the unxpected does appear, it can be characterized in detail and its nature can be known as soon as possible. Thus a sound base for correct prediction of the therapeutic benefits and hazards of the new substance can be established.

In this respect I believe that the interests of the regulatory authority coincide with those of the manufacturer. Quality in toxicology derives from sound general scientific principles.

REFERENCES

1. Duncan, W. A. M., (1963). *Excerpt. Med.* (I.C.S.) **73**; 67
2. Friis, C. W. (1966). Forsøgsdyrkunskap, Jydsk Teknologisk Institut, Aathus
3. Roe and Tucker (1974). *Excerpt. Med. Int. Congr. Ser.* **311**; 171
4. *Ronden 15* (1972), 233
5. Registration of pharmaceutical specialities, *National Board of Health and Welfare*, **31** (April 1974)
6. Federal Register, Vol. 41, No 225 November 19, (1976), 51206–51230

3

The seven deadly sins— a U.K. view
John P. Griffin

Toxicological evaluation of hazard is required on any new chemical entity to which man may be exposed, or any novel form of exposure to an already known substance. Exposure to this chemical entity may be as:

(i) A medicinal substance administered therapeutically to man by any route, e.g. orally, per rectum, topically, intravenously, intramuscularly, subcutaneously or by inhalation.

(ii) A veterinary medicinal substance present as a residue in meat, eggs or milk, e.g. steroidal hormone, oestrogen, veterinary therapeutic substance or growth promoter.

(iii) An agricultural residue in vegetable material, pesticide or selective herbicide.

(iv) A food colour or preservative.

(v) An environmental pollutant in air or water or even in food e.g. mercury in tuna fish or cadmium in shell fish.

(vi) As an industrial exposure, e.g. vinyl chloride.

It is apparent therefore that the toxicological evaluation of the chemical entity has to be tailored to be relevant to the specific problem, and appropriate to the nature of the exposure of man to the chemical. Such a study must therefore take into account whether this exposure is intentional or accidental, acute or chronic, the route of exposure, and whether the exposure is likely to be that of the single new chemical substance alone, or in combination with other agents, and whether any particular subsection of the population might be at specific risk, e.g. the developing fetus, and also includes the long term effects of the chemical in terms of carcinogenesis and mutagenesis.

In its broadest sense the toxicological evaluation of an agent covers single dose LD_{50}, short and long term toxicity studies, carcinogenicity studies and teratological and fertility studies. The guidelines indicating the current thinking of various drug regulatory authorities on these studies have been published and there is a wealth of literature covering the principles of all aspects of toxicology evaluation (e.g. Ballantyne, 1977). Therefore it is unnecessary to reiterate them here and it is intended to consider only some of those factors that affect the quality or integrity of toxicity studies. The burden of conducting a satisfactory toxicological evaluation of any new chemical entity is considerable both in terms of time and of financial committment. It is therefore a matter of very considerable concern that some toxicity studies that are initiated are poorly designed, or are inappropriate with respect to evaluating the agent in the manner in which human exposure occurs, or are not designed to generate the maximum amount of relevant information that could be obtained for little extra burden in time or cost. Other studies are seen which are (i) poorly conducted (ii) inaccurately recorded and (iii) incompletely documented.

Toxicity studies that are:

(i) defective in protocol design

(ii) inappropriate and irrelevant

(iii) conducted in a slovenly manner so that a maximum amount of useful data are not generated

(iv) dishonestly reported

are a waste of money, a useless expenditure of valuable time and an unjustifiable use of experimental animals. The use of large numbers of experimental animals in toxicological testing is a justifiable cause of considerable public concern. Whereas there is a need to use animals in safety evaluation studies there is a moral responsibility that such studies as are necessary should be conducted with high standards of integrity, using protocols designed to obtain the maximum amount of scientifically valid information from as economical a use of animals as possible. (The use of inadequately small numbers of animals in studies that eventually have to be repeated with adequate numbers is not economical use of animals, time or money).

In November 1976 the Food and Drugs Administration published in the Federal Register 'Proposed Regulations for Good Laboratory Practice'. The circumstances giving rise to a need for such regulations are clearly stated and are similar to many of the problems referred to above, and therefore the British Licensing Authority is in agreement with the FDA on the definition of the problem that exists on the quality of toxicity studies.

Any attempt to improve the quality of toxicological studies is laudable. However, any regulation introduced should not interfere with needed flexibility of laboratory operation, it should not stifle the use of informed scientific judgement and it should not freeze protocol design nor preclude technological advance. It is hoped that the proposed Good Laboratory Practice Regulations will achieve their stated aims 'to ensure the quality and integrity of data' and will not 'inhibit scientific study or burden laboratories with unnecessary or inappropriate requirements'. It is debatable whether good scientific work, in this case toxicological evaluation, can be achieved solely by regulation. It is to be hoped that such regulation will not have a limiting effect on the introduction of new techniques of toxicological evaluation. What is certain is that regulation or legislation can never ensure scientific integrity and honesty or overcome ignorance. The presence of such regulations only acknowledges the existence of these problems but is not necessarily a solution to them. (It may of course result in the employment of better trained personnel, but these are in short supply). It is a major problem that, when one has the task of criticising standards of toxicological studies, those who conduct poor quality investigations do not attend the lecture or symposium and all too often the most vitriolic remarks are heard by the innocent, not those at whom they are aimed. However, since none of us is perfect all the time we can all learn from the errors of others even if these errors are (i) uncommon or bizarre; (ii) have occurred by accident through mistake or ignorance and (iii) have been produced with an intent to mislead. The deficiencies in toxicological studies submitted to the Committee on Safety of Medicines can be considered under the headings of the following Seven Deadly Sins.

(1) Sins of ignorance or idiocy.

(2) Sins of omission.

(3) Sins of commission.

(4) Incompetence of personnel and sloth.

(5) Suppression of facts, distortion of the truth, and the drawing of misleading conclusions.

(6) Deliberate falsification of results.

(7) Misuse of Statistics.

1. SINS OF IGNORANCE

The sin of ignorance is usually immediately apparent on inspection of the protocol for the study, and the depth of ignorance becomes progressively more apparent as assessment of the study proceeds. The fundamental problem here is that of the quality or inexperience of the personnel responsible for conducting the study. In this category are found the people who cry aloud for regulatory authorities to design protocols which should be rigidly and slavishly followed; they are unable to design their own studies to the particular requirements of their specific problem. This group of people paradoxically are the first to complain of lack of flexibility when specific toxicity studies are requested.

There is one specific sin of ignorance which is voluntary, and does not relate to the quality or inexperience of the investigator and unfortunately is frequently seen and is deplorable. It is due to the not uncommon practice in which a manufacturer requests toxicological studies to be conducted on a compound by a contract research organisation and expects the study to be designed and conducted without disclosure of the chemical structure and pharmacological activities of the compound. Any toxicologist who enters into design and conduct of a study in these depths of voluntary ignorance, as I am aware that some toxicologists have done, is stepping into the realms of the sin of scientific idiocy.

This type of idiocy can be exemplified by this true story. A toxicologist was conducting a study in beagle dogs in a well designed animal house which had a long central corridor. Down this corridor the animal house staff were prone to practice their golf shots. During this relaxation some of the test animals ate the golf balls. The toxicologist therefore added a new group of control beagles to his study which he fed on golf balls. The brand of golf balls was not stated, and there was nothing to indicate in the report that the brand and quality of the golf balls used was the same as the test animals had ingested accidentally or even that the brand and quality of the golf balls fed to the control group was uniform throughout the study. The fact that the control group dosed on golf balls showed no untoward effect does not preclude the interpretation that the toxicity seen in the top dose can have been due to an interaction between the drug and the golf balls.

The same level of ignorance of drug quality is frequently demonstrated (i) in the context of toxicity tests which are conducted on material on which there is no batch analysis or specification of impurity pattern available (ii) toxicity tests conducted on two or more batches of material with greatly different specifications being used in the study and given to different dose level groups simultaneously or administered sequentially. (iii) particle size of drug administered orally is seldom known to the toxicologist.

In many cases the ignorance of the toxicologist on the specification of the drug exists merely because he does not ask to know, or realize that this knowledge may be essential to the conduct of the study. Quality control of the animals used in toxicity studies is often overlooked and the most valuable studies are those conducted in circumstances where a great deal of information on the background and pedigree of that strain of mouse, rat and rabbit or beagle colony is available. Studies where animals are 'bought in' from an animal dealer where little background data are available often leads to problems. This is a particular cause for concern in many primate studies where the animals used are newly arrived from the jungle. Many of these problems of ignorance arise because there are inadequate training programmes for toxicologists, and most toxicologists learn by their mistakes, or, if they are lucky, from other people's. There is a definite need for more formal training programmes in the field to produce competent toxicologists.

2. SINS OF OMISSION—'Leaving undone those things that ought to have been done'

In toxicology sins of omission arise from three main root causes. Firstly those which stem from ignorance, secondly those which stem from parsimony and commercial greed, and lastly those which stem from laziness on the part of the protocol designer, or the laboratory staff.

Those sins of omission which stem from parsimony are all based on the philosophy of 'How little need we do to satisfy the regulatory authority'? It is natural that in a field where investigation is expensive a concern should be expressed that time and money should be used as economically as posssible and the attitude of the responsible toxicologist must be 'This compound must be evaluated thoroughly and the study must be designed to enable this to be achieved quickly and economically'. With these sentiments none can quarrel. In those studies conducted with the 'how little can we get away with' philosophy, almost every fault has been committed in one study or another.

(i) Inadequate recording of data.

(ii) Inadequate or absent control animals.

(iii) Inadequate numbers of dosage levels used.

(iv) Inadequate numbers of animals per dosed group.

(v) No predosing data from any of the experimental animals.

(vi) Inadequate data on dose levels given, and appropriate selection of dosage levels.

(vii) No monitoring or inadequate monitoring during the study.

(viii) Inadequately detailed terminal monitoring and autopsy report.

The worst of these basic sins of omission were referred to in Proposed Regulations for Good Laboratory Practice and it is sad that this has to be said in an age where education stresses accurate recording of scientific data in laboratory note books from the lowest forms of secondary education. Yet these Proposed Regulations stress the need to keep satisfactory records again and again, and in such elementary detail—'to be recorded promptly, and accurately in ink in bound books. . . .' 'All data entries are to be dated and signed. . . .' 'Any changes made are to be made in such a manner as not to obscure the original entry. . . .' The words used are almost identical with the instructions given to any science class in the first form in a comprehensive school. Yet there is no doubt that in many cases record keeping is poor.

In the realm of economical handling of toxicity studies it may be necessary to give the drug in the diet rather than daily administration by gavage and this is justifiable, in for example, a two year carcinogenicity study in the rat. A frequent and major sin of omission in this context is that no attempt is made to determine that the animal is actually getting the amount of drug from the diet, in mg/kg/day, that was intended. If this dose level is actually achieved initially it is rare that attempts are made to adjust the 'food/drug mix' at intervals during the study to take account of the changing food intake of the animals and thereby achieve a steady drug intake. Even more rarely are the actual plasma levels monitored in these drugs in the diet studies despite the very frequent uncertainty of drug intake.

At this point it is valuable to mention that in any toxicological study it is desirable to measure the plasma levels of the drug (if this is technically possible) at least at two points in time in the duration of the study, to determine whether cumulation or enzyme induction occurs. Metabolic studies are all too often omitted from toxicological testing or treated separately in some peculiar demarcation dispute.

In order to interpret toxicological findings it is necessary to know whether the species and strain used handle the compound similarly to man. The metabolic handling of a compound may be markedly different between species and this is generally accepted, but cases could be cited where metabolic differences between strains of rat have been demonstrated. It is a wise principle to conduct metabolic studies on the same species and strain of animal to be used in toxicity studies.

Probably many drugs have been discarded on the basis of toxicological findings in animals, where a little more effort would have shown that they were irrelevant findings as far as the therapeutic use of the drug in man was concerned. Discarding a useful drug for the wrong reasons is not in the manufacturer's or public's best interest. A particularly common sin of omission is demonstrated in those slovenly studies in which autopsies have been conducted only on the top dose group on the basis that if these animals do not reveal any pathological change, then it is likely that there will be no changes at the lower dose levels.

It is worth dwelling a little on this approach to conducting autopsies on the top dose group of animals only. If the study has been properly designed it should have been conducted using a top dose that would reveal the target organ toxicity and provide a guide as to the margin of safety of the compound and what its toxicity might be in man. (The purpose of toxicity studies are not to provide a marketing organisation with a set of negative findings on which to build a glowing testimonial of safety). If this philosophy of conducting a toxicity study on a therapeutic substance is to demonstrate the target organ it will be appreciated that autopsy at the lower dose levels is necessary if a 'no-effect' dose is to be demonstrated. This undesirable practice of examining only control and high dose level animals is also referred to, and commented on adversely in proposed 'Good Laboratory Practice Regulations' (1976). It is obviously a common sin of omission on both sides of the Atlantic.

It must be equally apparent that predosing control measurements should be made on animals before commencing the study. If abnormalities of haematology, serum biochemistry, urinalysis or pathology are found terminally what assurance is there that the changes are, or are not drug induced if there are no predose control data for that animal.

3. SINS OF COMMISSION

Failure to adhere to the protocol is the principle sin of commission and transgresses the first law of good laboratory practice. The FDA in their proposed regulations for 'Good Laboratry Practice' have put their finger on this problem and they write 'technical personnel were un-aware of the importance of protocol adherence'. Perhaps this sin of commission may really be a sin of ignorance, in many cases it appears to be a sin of sloth; of taking short cuts and omitting to conduct investigations in the hope that they will not be noticed. In other cases intentional sins of commission may be committed in non adherence to the protocol and this may be done for many reasons, few of which are justifiable.

4. INCOMPETENCE OF STAFF—INADEQUATE TRAINING AND SLOTH

To those who design protocols and interpret toxicological data these studies can be interesting, but to those technicians who have the day to day running of the study they can be tediously repetitive and boring. Boredom leads to loss of interest, careless work and sloth. In this boredom factor lies a major problem in the quality of these studies, and the longer the duration of the study the greater this quality problem.

There is a definite need to involve, motivate and maintain the interest of all animal house and technical staff throughout the duration of studies. Involvement can be maintained if the protocols are imaginative rather than routine and repetitive, and understanding of the requirements of the specific protocol are conveyed to all levels of the investigatory staff. Toxicological studies using a standard unimaginative protocol irrespective of the particular problem the drug imposes is a rapid way to end up with bored and slothful laboratory staff as well as a possibly inappropri-ate study. Where human use is intended to be for 7 days per week, dosing on less than this number of days per week in animal studies is unacceptable. It has been noted that some applicants accept this in principle but fail to carry it out in practice. In order to administer the correct amount of drug per week, some units have adopted bizarre dosing regimes. The most extreme case is probably the 5 day per week dosing with double doses given on Monday and Friday.

5. SUPPRESSION OF FACTS AND DISTORTION OF THE TRUTH

The problem of suppression of facts is widespread. A typical case occurs along the following lines; a toxicological study has been conducted and gives an equivocal result, or a result unfavourable to the product. A second study is conducted and at times even a third in which the dose levels are adjusted or the protocols modified in such a way that eventually a result favourable to the applicants product is obtained. Only the result favourable to the applicant's product is submitted to the regulatory authority. A second example of suppression of a similar type can be cited. Microscopical examinations of histopathological slides may be made by more than one pathologist each of whom may have come to different conclusions, yet only the conclusions favourable to the drug are submitted to the regulatory authority. On one occcasion where such a situation had been detected the applicant with a dismissive gesture said 'that investigator gives the wrong results, we will not use him again'. (This attitude reveals the commercial pressure that can be brought to bear on an investigator by the threat of loss of future work).

A recent example of the suppression of truth can be cited from H.E.W. News October 14, 1976 in which in a 2 year rat carcinogenicity study on Naproxyn the existence of tumours in some animals was not reported, and the study was also alleged to contain misstatements and to have omitted essential information. It is also alleged that many of the 47 animals described in one log book as having been destroyed without examination were reported in another log book as having been examined. (No complete set of records was maintained for any single animal among the 160 that began the study).

Fabrication of results is not as common in toxicity studies as it is at the clinical trial stage. In the latter fabrication may be minor and involve the invention of a few measurements (usually invented to make up for the gaps in a series of measurements on a patient, possibly due to a patient's non-attendance at a clinic) or in a few well known cases has even involved the invention of the patients from whom the imaginary data are derived. If and when fabrication occurs in the toxicology laboratory it appears to be limited to the invention of a few observations rather than a major constructive exercise. Often these inventions are done at a technician level to conceal the loss of a blood sample that has been dropped, spilled or mislaid. On a few occasions the fabrication may arise because an *actual* measurement is rejected as unacceptable or inconvenient.

6. DELIBERATE FALSIFICATION OF RESULTS— TERMINOLOGICAL INEXACTITUDE

Public concern about falsification in scientific research is growing since the allegations published in the Sunday Times (24th October 1976) that Sir Cyril Burt faked results in some studies. The responses to a New Scientist questionnaire published under the title 'Cheating in Science' (St. James-Roberts 1976) indicates that the problem is not rare among scientists in any discipline. Among the 204 returned questionnaires in this report 92% of the respondents had some sort of direct or indirect experience of intentional bias, 8% did not. (Examined critically it will be appreciated that the first question of the questionnaire 'Does intentional bias warrant investigation'? would obviously itself intentionally or un-intentionally elicit replies in the form of returned questionnaires from the sector of the population who would tend to answer this question in the affirmative—in fact 90% did indicating a strongly biased population sample, i.e. the questionnaire itself is an example of poor experiment design). The preamble also suggested to the reader that if they knew of a problem then the questionnaire should be completed i.e. the study was designed to give a biased answer throughout. The lesson that we can learn from this study is that toxicity studies too can be so designed by a fundamental bias in the protocol that any answer required can be obtained. Falsification of the true toxicity of a product can be obtained not only by fabrication of the results but also by building bias into the study design so that even if the protocol is honestly conducted a false answer is obtained.

7. MISUSE OF STATISTICS

The proper use of statistical techniques in the analysis of toxicological data is not being condemned. Statistical analysis is a valuable and legitimate tool to enable a mass of toxicological data to be analysed and condensed into a comprehensible and interpretable form and to give an indication of the probability of any finding occurring by chance.

The application of sophisticated statistical methods to dubious data derived from a poor quality toxicological experiment in the hope of giving it an air of respectability is not uncommon. A statistical analysis is only as useful as the data being analysed, and bogus data is not improved by a high quality statistician, spending hours on its presentation in a manner designed to obscure its shortcomings. Statistics are also used to obscure unwanted data on the drug or substance under test. A case can be cited where some dramatic falls in haemoglobin of the order of 3–4 g/100 ml in two animals were attempted to be hidden by presenting the haematological data as means and standards errors and commenting in the text that overall the mean haemoglobin levels were only slightly reduced when before and after treatment values were compared. It is often overlooked that toxicological studies are conducted on relatively small numbers of animals and consequently the data derived from each animal are too valuable to be fudged to conceal findings in this way.

INTERPRETATION OF TOXICOLOGICAL STUDIES

Collection of data from well designed studies does not necessarily mean that the safety of the product will be ensured. Any toxicological study needs to be interpreted and it must be interpreted in the light of the proposed clinical use taking into account the dose levels that are to be used therapeutically, the likely duration of clinical use and a careful assessment of the benefit-to-risk factor in the patients in whom the drug is to be used. This assessment needs even greater experience than that required to design the study.

TOXICOLOGY IN MAN

The most valuable toxicological data is obtained from the administration of the drug to man, and it is necessary that proper monitoring of all patients treated in early clinical trials is obtained. The quality of many clinical trials is good with respect to assessment of the efficacy of drugs but may be deficient in safety evaluation. It must be remembered that many serious adverse reactions to drugs cannot be predicted from animal studies, for example methyl-dopa-induced haemolytic anaemia, the effects of phenylbutazone on bone marrow or more recently the oculo-cutaneous syndrome of Practolol. Proper haematological and bio-chemical monitoring and recording of adverse reactions from patients treated during clinical trials and after marketing provide toxicological data of the greatest value, far exceeding that derived from animal studies. Unfortunately the quality and honesty of some clinical studies are open to the same criticisms made of animal toxicology. This is more serious in its consequences, since animal toxicology has its limitations, in its relevance to man, but a human adverse reaction to a drug is undoubtedly relevant.

CONCLUSIONS

It would be simplistic to suppose that a code of Good Laboratory Practice will eliminate all the problems that detract from the quality and integrity of toxicological work. It must be appreciated that a code of laboratory practice and an inspection system such as the FDA propose could if properly exploited achieve (i) better laboratory conditions for the animals used in the study (ii) ensure better laboratory data recording and (iii) over a period of years slowly perform an educative role. An achievement of these ends should be welcomed.

However the expenditure involved in such a system may not be cost effective. On the whole the Committee on Safety of Medicines has been satisfied that by a very detailed examination of data presented many of the Seven Deadly Sins can be detected. However, they and the Licensing Authority, are not complacent. At the present time it is felt that the power to reject applications for licensing of new drugs if the toxicity studies are inadequate, inappropriate or of poor quality is adequate enforcement power. There is no doubt that the Licensing Authority has played and should continue to play an educative role, and aim to improve standards in all aspects of work required for drug registration. One method employed is by the issue of guidelines as to what requirements are likely to be. Notes for Guidance on Reproduction Studies MAL 36 which was prepared by the Committee on Safety of Medicines in conjunction with experts from academic departments, and industry, for the Licensing Authority is such an example. It is shortly hoped to issue 'Guidelines on Carcinogenicity Testing of Drugs' which are being drawn up by a Committee on Safety of Medicines Working Party including experts from industry and university departments. The Secretariat are always available to offer advice on specific problems.

Over the years even since the original Committee on Safety of Drugs was established in 1963 a dialogue with industry and appropriate experts has been conducted. In the coming year it is intended that the Medicines Division of the DHSS will hold a 'teach-in' for the industry. This will be a more elaborate venture into the educative role, that already some of the Scandinavian Regulatory Authorities have conducted with considerable success.

Finally it must be concluded that may be there are three basic problems of Quality Control in Toxicology (i) ignorance (ii) human error (iii) humanity's failing through what a theologian might call 'original sin'. The solutions to these problems are (i) education (ii) quality control systems designed to overcome or minimize errors (iii) the appointment of competent staff with integrity.

REFERENCES

1. Ballantyne, B. (Ed.) (1977). *Current Approaches in Toxicology.* (Bristol: J. Wright)
2. Food and Drug Administration (1976). Proposed regulations for Good Laboratory Practice. *Federal Register Part II* Nov. 19, 1976

SECTION TWO

The Industrial Viewpoint
Chairman:
G. E. Paget

4

Some possible effects of the introduction of quality control systems on the discovery and safety evaluation of drugs
B. J. Leonard

INTRODUCTION

Anyone who has written a paper for a scientific journal must be aware that errors in the recording of experimental work can be discovered right up to the reading of the last editorial proof. With care, errors can be reduced to a minimum but rarely, if ever, excluded entirely. The regulatory function of a Government-appointed agency is to determine the predicted safety of a new drug on the evidence of pre-clinical toxicity work presented to them. This totally impartial assessment is generally recognized to be in the public interest and I know of no toxicologist in industry who would disagree with this. Regulatory bodies need accurate data and it is our task to make sure we provide them with such data. Therefore, no one will disagree with Good Laboratory Practice: the only debate is the best means of achieving this, and we should keep constantly in our minds what should be the main purpose of our activities—which is, the introduction of significant new drugs of benefit to medicine, following an appraisal of the benefits and risks.

I joined ICI Pharmaceuticals Division 10 years ago following post graduate appointments in London and Manchester Universities and consultant posts in the Health Service. I was attracted to ICI Pharmaceuticals Division because of an interest in experimental pathology generated early in my career. I made a conscious choice between a pharmaceutical company and a university department because I found that the basic requirements for experimental pathology were considerably greater in the ICI research laboratories.

BASIC REQUIREMENTS FOR QUALITY ASSURANCE

Before discussing the organisation of chronic toxicity tests and the quality control schemes which are now in operation in ICI Pharmaceuticals Division, I should like to state some of the basic requirements for quality assurance in any research investigation.

1. A supply of healthy animals

This is a pre-requisite to any experimental pathology work. ICI was one of the pioneers in the development of an SPF colony of rats and mice. In any study in which animals are required to survive for many months or years, it is most important that the naturally occurring diseases are kept to a minimum, and a knowledge built up of spontaneous diseases in the colony of animals used. Today, a long term carcinogenic study requires approximately 700 mice and 700 rats, or other common laboratory animals, for evaluation, and therefore a supply of healthy animals of known age, of known pedigree, and known spontaneous disease rate becomes more critical with these very large tests. I know of no university department in this country which could supply and accommodate animals over a two year period for the evaluation of compounds for carcinogenic potential.

Most regulatory bodies state that there should be a choice of animal species for chronic toxicity tests and that attempts should be made to try and match animal species for similarities of metabolic disposition with man. In other words, for the non-rodent species we require a species other than the dog. ICI has its own breeding colony of Beagles, and over the last eight or nine years we have developed a breeding colony of marmosets (*Calithrax jacchus*) to provide us with an alternative choice.

2. Animal accommodation

It is necessary to house different species separately and to maintain high standards of animal husbandry throughout the studies. In chronic and carcinogenic studies we allow very limited access to animal rooms housing these studies.

3. Recruitment, training, environment

The next basic requirements are the recruitment and the training of staff in the disciplines involved in toxicity evaluation. This requires the creation of the right environment for innovative scientific work. At least a portion of their time should be free to carry out research into better methods of assessing animal toxicity which are relevant to human safety.

4. Quality control systems

These will be discussed in detail later: sufficient to say that these have been developing and evolving over many years.

5. Storage and retrieval of information

A good system of information storage and retrieval is essential in studies of such magnitude. Let us take, for example, a current carcinogenic study in two species. This will require the accurate recording of observations and findings in some 1400 animals and will result in a minimum of 56 000 pieces of tissue to be examined microscopically. A team of four scientists, in our experience, can carry out 15 to 20 post-mortems per day. This means that an evaluation of one compound will require the full-time commitment of such a team for 13 weeks.

Clearly all these pieces of information have to be recorded accurately and I believe the finding of some discrepancy in one or two pieces of information in a total of 56 000 separate tissue examinations should not necessarily be interpreted as dishonest practice. One should try and achieve 100% accuracy but have sufficient humility to accept that human beings, being what they are, are bound to make some errors.

6. Trained staff

With such extensive investigations it is an essential part of management to ensure that there are sufficient trained staff in the various disciplines to carry out the tests in a methodical manner without imposing rigid time constraints or pressures on them.

7. Reporting

It is also important that the responsible scientist should feel confident that his opinions are expressed correctly and that the pathologist who actually undertakes the study should have sufficient experience to write the report on such a study.

8. Communications

Because any long-term study requires a multi-disciplinary team one of the important requirements to any quality assurance is to achieve efficient communications so that all information relevant to a particular study is communicated to the staff involved. Means by which this end can be achieved will be discussed later in this paper.

CLINICAL PATHOLOGY

The improvement in quality control in toxicity studies has been a continuously evolving process, for example, in clinical pathology we introduced a quality control scheme as early as 1972. I am indebted to Dr John Fowler who is in charge of our clinical pathology for advice on the comments which I am about to make.

Quality control in clinical pathology is achieved by the actions of laboratory staff and is based upon various monitoring studies. These monitoring studies include the frequent analysis of known and unknown samples from a variety of sources 'in house', but mainly 'proprietary'. In addition, at regular intervals, samples are received from interlaboratory exchange schemes run by Wellcome ('Welcomtrol'), and the NHS (National Quality Control Scheme, Queen Elizabeth Hospital, Birmingham) of which we are members.

The quality monitoring analysis results, upon which quality control is based, will constitute from 10% to 30% of laboratory workload, depending on the known inherent stability and reliability of the various instruments. In our laboratory, which generates nearly 200 000 requested results each year, an additional quality control workload is generated, amounting to 30 000 quality monitoring data points. This workload justified the appointment of a Quality Control Officer in 1973 and until early 1976 the Quality Control Officer's job was the priority task of the individual and took up to 65% to 70% of her time. The remainder of her time was devoted to the speciality expertise of the individual. In future, large parts of the job are going on to the computer which will also increase the degree of control exerted and minimize the chances of issuing unsatisfactory results. The Quality Control Officer is responsible to the Section Head of the Clinical Toxicology Section. The staff in the Section provide documentation of the standards of work for the Section and the Division quality assurance staff. The objective of quality control in clinical pathology, as in other analytically based disciplines, is the attainment of a high degree of accuracy* with good precision.†

The attainment of the objective of accuracy enables the results of the laboratory to be compared with other laboratories on any given day and also with the results obtained on other occasions, for example, those of 1976 with those of 1973. The attainment of the objective of precision is essential for any analytical method: the precision of results governs the confidence with which predictions can be made.

The quality monitoring activity is stepped up until minute deviations in analytical precision and accuracy can be detected. As long as pre-determined limits of acceptable accuracy and precision are obtained, a method is deemed 'in control'. If these limits are exceeded in a systematic or regular or frequent manner, the method is deemed 'out of control'. No results are released unless methods are 'in control'. The limits of acceptability must be chosen so that a method becomes technically 'out of control' before the error contributed by the analysis assumes biological significance.

Differences between sample analysis results arise in two categories:

1. biological variation
2. laboratory variation.

Biological variation is due to differences between individuals, toxic effects of chemicals, manifestation of disease, stress, circadian rhythm, husbandry, sex, age, etc. In order to minimize the contribution of factors other than those that are drug-induced, we choose to work with animals under the following conditions:

(a) the animals are from a stable genetic pool (e.g. ABU rats)

(b) they are substantially disease free

(c) they are not exposed to dietary, or environmental, contaminants

(d) they are housed at even day length and temperature, without undue noise, etc.

(e) they are tended and regularly handled by the same staff

* 'Accuracy' is defined as achievement of the 'true' value.
† 'Precision' is defined as the 'reproducibility' of the values from replicate analyses.

(f) they are dosed and sampled at the same time each day by the same staff.

By stabilizing the various factors above, it is anticipated that differences between normal and abnormal animals will be detected. This is conventionally achieved by reference to so-called 'normal ranges'. Normal ranges are established by pooling results from large numbers of control (i.e. untreated) animals gathered over long periods. Recently, the conventional normal range concept is falling from favour and clinical pathologists in toxicology favour the establishment of 'range of expected values' before treatment begins. The invalidity of the 'normal' range concept arises from failure to stabilize truly and control the various features of biological variation described, and also from the contribution of laboratory variation, which will arise despite the application of conventional quality control procedures. The contribution of the laboratory to overall variation arises from two major sources:

1. the restraint, sampling, sample collection, and sample preparation procedures.

2. the storage of samples, accuracy and precision of the analytical method and documentation procedures.

In contrast with animal toxicology monitoring, human clinical pathology is based on large volumes of sample taken with relatively little stress. Even so, the value of 'screening exercises' is not proven and provides a basis for continuing discussion. In human medicine, the value of sequential monitoring is, however, clear, and unambiguous results are obtained from laboratories with good quality control performances, good staff, and modern automated equipment.

The conventional quality control procedures of the animal clinical pathology laboratory are insufficient unless backed by rigidly standardized laboratory practices in the animal house. Dr Fowler and his staff have shown conclusively that a large component of the so-called 'normal range' in small laboratory animals, such as rat or marmoset, arises from the sampling procedure. This must be a prime consideration in evaluation of laboratory variation and the contribution of the analytical component. The contribution of laboratory variation to total variation must be understood if biological variation (the usual reason for analysis requests) is to be quantified.

The clinical chemistry laboratory of Safety of Medicines Department has shown a consistent improvement in its precision index since we became members of the National Quality Control Scheme (NQCS) in 1973. The performance places the laboratory in the top 5% of all laboratories in the UK. Much of the equipment has been identified and purchased on the basis of published quality control data obtained from our membership of the NQCS. Samples are analysed 'blind' by the laboratory and attainment of a high quality of work has become a major objective for laboratory staff. They are able to take pride in the speed, accuracy and precision of the laboratory's work. I have gone into the quality control of clinical pathology in some detail because it does highlight the fact that this type of work ante-dates any Regulatory demand for quality control, and we are conscious of the commitment of people in this laboratory to producing reliable results on which we can make decisions.

CENTRAL RECORDS AND INTERNAL QUALITY CONTROL UNIT

When I joined the Pharmaceuticals Division in 1967 the toxicity work was carried out in the Biology Department and at that time there were about thirty staff in the Toxicology Section. With the ever increasing size and complexity of pre-clinical toxicity tests senior management decided in 1971 to establish a separate department to be known as Safety of Medicines Department. This department has grown rapidly and the manpower figure for 1977 is 183 and the projected manpower figure for 1980 is 250.

When the department was formed we made a new appointment of Scientific Secretary to the department and recruited an experienced scientist. His work remit was to check all documents for accuracy, and assist in the collection, storage and retrieval of all relevant information contained in submissions to Regulatory bodies. With the introduction of the Good Laboratory Practice procedures we have enlarged the remit of the Scientific Secretary and integrated several work Units under the Central Records and Internal Quality Control Unit whose main tasks include the collection and maintenance of data on studies, both completed and current, and the management of the current archives and the permanent archives of the safety evaluation studies. All raw data and reports on drugs under development are kept in the current archives until NDA approval, when all information related to that drug is transferred to the permanent archives for storage.

The Central Records Unit provides a quality assurance function on all incoming data and houses all protocols on development drugs: it maintains a master schedule of all tests carried out on each drug, and all raw data related to the various experiments. The microscope slides and paraffin blocks are housed in the archives and we have found it a useful practice to transfer paraffin blocks to the archives as soon as

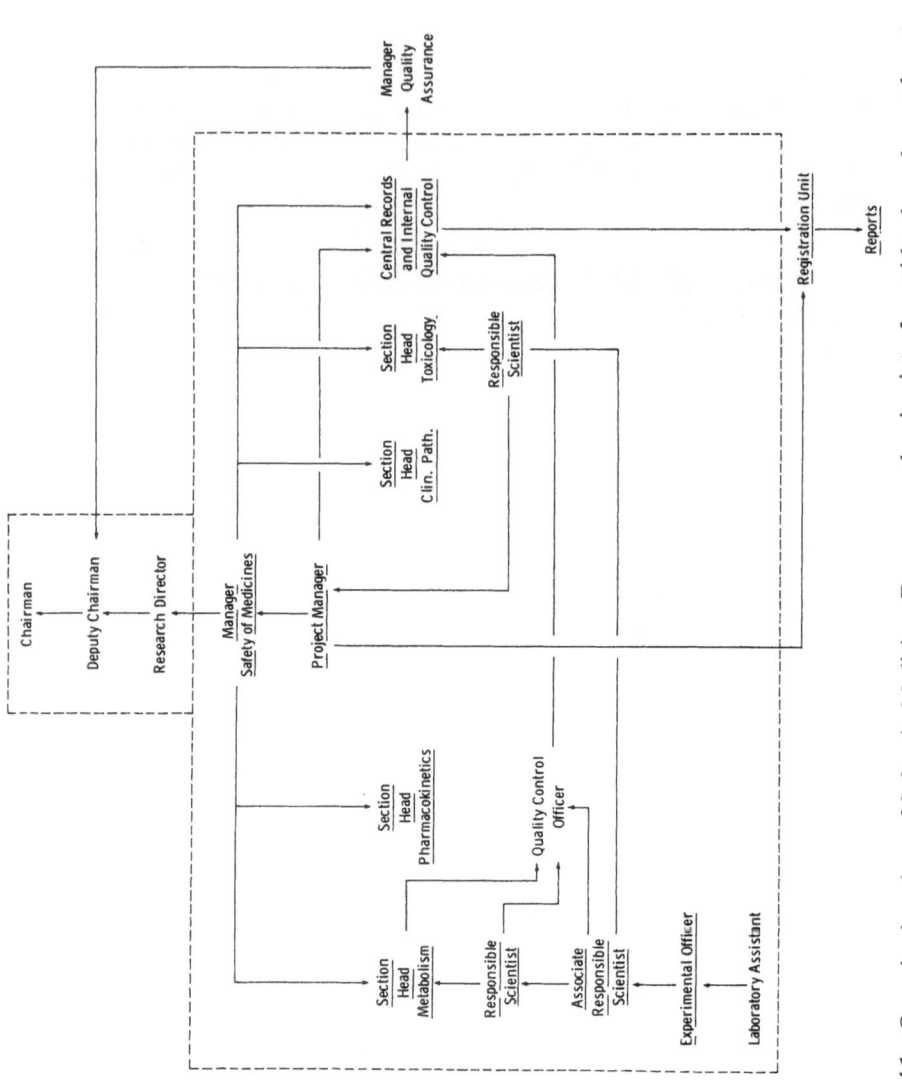

Figure 4.1 Organization chart of Safety in Medicines Department showing interface with other relevant departments

sections are prepared to avoid any loss or misplacement of the specimens. In addition, this Unit maintains a register which records the receipt and issue of drugs and the quality control studies which are carried out on drugs used in the toxicity studies. Further, because of the risks of fire and theft of records, we routinely microfilm all our records and house these films in a separate establishment.

The Central Records Unit is at present helped in monitoring the toxicity studies by computers. We have tried at ICI to make computers play an active role in monitoring the quality of data, to control experiments, to store, retrieve, analyse, and report data, and to aid management. Vast amounts of different types of data are generated during toxicity studies. Animals are weighed weekly, biochemical and haematological studies are performed at several different times during the trial, and organs are weighed and examined pathologically at the end of the trial.

The computer systems we operate were designed as an aid to the collection, storage, analysis and reporting of the data, and the management of the different laboratories concerned. The following data is at present in-put, on-line, through visual display units:

body weights	diet intakes	clinical chemistry
urine analysis	organ weights	clinical observations
dose changes	all routine (including validation) checks	

The scientist is immediately alerted to any potential errors, e.g. body weights are compared with expected values for the age, sex and species of the animal, thus minimizing weighing and typing errors. Statistics on body weights and diet intake are calculated automatically and summary data can be displayed immediately on the visual display unit. The systems have reduced the clerical work involved in data collection and analysis, helped in the evaluation of the progress of the trial, and improved the accuracy of the final report. The computers are used in resources allocation, work flow management, identification of potential toxicological effects, automatic report generation, and availability of data.

'EXTERNAL' QUALITY CONTROL ORGANIZATION

In addition to the Internal Quality Control Unit about which I have spoken, we have now established a new position in the Division of Scientific Practices Manager and he responds directly to the Deputy Chairman. This external quality assurance role will cover the safety evaluation work of the Safety of Medicines Department of Pharmaceuticals Division, and also the toxicity work carried out in the Central Toxicology Laboratories of ICI which deals with the safety evaluation of non-medicinals. This new Unit will have essentially an external audit role over all the safety evaluation studies carried out by Safety of Medicines Department and will be seen to have the advantage of independance of the management directly concerned with carrying out the pre-clinical toxicity studies.

Project teams

The research and discovery of new drugs in the Pharmaceutical Division is carried out principally by project teams. Each team is multi-disciplinary and consists of chemists, biologists, toxicologists, who have a research target such as anti-inflammatory drugs. These teams meet regularly under the chairmanship of the Research Director and all members of senior management in the Research Department. When a chemist synthesizes a new chemical and the biologist discovers an interesting pharmacological property of the chemical this will be discussed by the project team and the Research Director will decide whether to submit the chemical to a research toxicity study. In Safety of Medicines Department we have a Research Toxicity Unit whose role it is to investigate compounds submitted by the project teams as described. The purpose of research toxicity is to discover at an early stage in the life of a potential drug whether there is any obvious unacceptable toxicity which preclude further development. The tests include dosing for up to one month in two species, and mutagenicity studies, etc. The investigations performed in the Research Toxicity Unit include tests, for example, mutagenic tests, which are not required by Regulatory bodies. Provided the drug does not show any obvious unacceptable toxicity, the biologist and pharmacologist will carry out further work identifying the pharmacological properties of the drug in more detail, and the Research Director will decide to set up a Development Team to monitor the further development of the drug which at this stage is designated a 'candidate' drug.

Development team

The Development Team consists of a chairman, who is a member of the Development Department, the biologist, biochemist and chemist, who were associated with the discovery of the drug; a member of the Clinical Research Department, who will eventually undertake clinical trials if the drug is declared safe in the pre-clinical toxicity tests; a member of the Formulation Section of Pharmaceutical Research Department, and members of Safety of Medicines Department. This department identifies a pathologist and a chemical toxicologist as members of the Development Team. Depending on the nature of the study, a pathologist or a biochemical toxicologist will be appointed as the Responsible Scientist for that particular candidate drug.

Responsible scientist

Detailed specialist studies, e.g. the pharmacological evaluation of chronically dosed animals, reproductive toxicology, metabolism and pharmacokinetic studies, will be under the supervision of a Responsible Scientist of the appropriate discipline. Where these disciplines play a supportive role in the chronic toxicity tests, then a number of Associate Scientists will ensure that the studies required by the Responsible Scientist in his protocols are correctly designed and implemented. On appointment, the Responsible Scientist must become familiar with all available information on the drug, e.g. the pharmacological properties, the dosage forms, and the results of the Research Toxicity studies.

The Responsible Scientist drafts protocols well in advance of the commencement of animal toxicity studies. The draft protocol is submitted to the Project Manager who will ensure that the proposed methods of testing fully evaluate the potential toxicity of the drug in order to satisfy both the requirements of international Regulatory standards and those of ICI. The Project Manager, in consultation with the management of the Safety of Medicines Department, will ensure that there are sufficient human, animal, and material resources to conduct the test satisfactorily, and will decide on the commencement date.

When the protocol has been approved by management the Responsible Scientist will ensure that it is followed exactly. In the event of any unforeseen circumstances which indicate that the protocol requires alteration, the amendments must be approved by the Project Manager and must be fully documented, and all those associated with the study must be informed in writing of such agreed change.

The Responsible Scientist must ensure that the dosed and control substances or mixture, or diet analysis, have been properly tested for identity, strength, quality, purity, stability and uniformity. He should alert management when the scheduled resources are not made available, and ensure that personnel clearly understand the function they are to perform, that they operate in a safe manner, and that all data are accurately recorded and verified. Any unforeseen circumstances which may affect the proper performance of the study, whether physical or due to the ill health of animals or personnel, should be reported to management and fully documented. All observations must be promptly and adequately recorded, and all applicable Good Laboratory Practice regulations followed. The Responsible Scientist must ensure that all raw data documentation and other information to be retained—protocols, specimens and final reports—are transferred to the archives during or at the close of the study.

When a chronic toxicity study has been completed and the Responsible Scientist, for example, a pathologist, has examined all the microscope slides prepared from the tissues removed from the animals, it is his responsibility to write the first draft report giving all the details of the study, the findings and his interpretation. The draft report will then be submitted to the Project Manager. If there is agreement on the interpretation of the findings in the chronic toxicity study then the draft report will be circulated to members of the Development Team and the management of Safety of Medicines Department. Any suggested alterations or amendments to the report, particularly if they concern interpretations of the findings, must be made in writing and recorded, and no changes to the report can be made without the agreement of the Responsible Scientist who conducted the study. In the event of disagreement on, for example, pathological interpretation, an external consultant would be asked for his opinion and the final report would record the opinions of all pathologists. The final report is sent to the Registration Section of Development Department and it is their responsibility to ensure that it is submitted to the appropriate Regulatory body.

Studies performed after a drug has been approved for sale are documented in a similar manner and headed 'Supplementary Report'. These reports are forwarded to the Registration Section whose responsibility it is to ensure that they are sent to the appropriate Regulatory body.

EFFECTS OF STATUTORY QUALITY CONTROL SYSTEMS

I have attempted to outline how we try to ensure that the quality control systems operate efficiently and that the results of our studies are accurately recorded with the Registration Section for reporting to the regulatory bodies. We recognize the need for good quality control supervision of all our toxicity studies and this has been developing over many years. I find myself very much in agreement with what is contained in the FDA publication on quality control of pre-clinical toxicity testing. However, I think there are limitations in the value of the present approach and I do not believe that it will necessarily lead to fewer adverse effects seen in patients who take drugs which have gone through such a procedure.

Those who have the responsibility of assessing the safety of candidate drugs by pre-clinical testing procedures in laboratory animals should recognize two distinct hurdles they have to jump: the first is to gain the approval of the regulatory body to conduct clinical trials on the drug. This is most important—but not nearly so important as what happens eventually to the drug when it is administered to thousands of patients after the drug has been approved for sale. In my opinion, most of the adverse effects seen in the clinical situation are due either to a particular test not having been done at all, or because the animal tests cannot predict the type of toxicity which is seen in the human. For example, the brain damage in babies following bathing in hexachlorophene was not suspected as a possibility due to the fact that animal experiments involving neo-natal bathing in rats and monkeys were not carried out; but no amount of animal testing could have predicted the toxicities we have seen in patients receiving chloramphenicol, methyldopa, etc. and these are all serious side effects. The toxicity tests were well-conducted and would have satisfied any inspection as outlined in the FDA quality control systems. Toxicologists can only evaluate the predicted safety of drugs in the human according to the best scientific knowledge available to them at the time. I believe that many of the pre-clinical toxicity tests which we use today are far too empirical and should be questioned. Real advancement in understanding of the pathogenesis of adverse effects will only come about through the ability of pharmaceutical industries to recruit innovative scientists to look into new methods of detecting toxicity. I cannot believe that the present proposed systems of inspections will be conducive to the recruitment of this type of scientist.

The extension of inspections to the research function in discovering drugs in my view would lead to a very considerably reduced number of new and important drugs. The conceptual scientist is exploring new territory and it would be quite impossible to write out a detailed protocol before the work is undertaken. In addition, the type of scientist who discovers a new significant drug is not the sort of person who works according to a 'cookery book' and very often he is unconventional in his approach. Indeed, it may be said that these very qualities are required for true discoveries. I should be very unhappy if I thought that this type of individual would be put off coming into industry where his talents could lead to important new discoveries. I feel that the inspection procedure should only start when a compound is declared a candidate drug and then, by all means, all the toxicity testing and other tests can be protocolled and inspected in the manner described in the FDA document. In general, the proposals place far too great an emphasis on numeracy and take too little account of the human element.

Let me give you examples:

1. At one time patients entering hospital who were thought to have anaemia had a haemoglobin and blood film taken and an experienced professional man, on the basis of the haemoglobin level and the appearance of the film microscopically examined, decided on what further investigations were required. This had the important advantage that it brought the professional man into intimate contact with the problem and avoided unnecessary investigations. Today, a patient entering hospital with a hernia or ingrowing toenails is likely to have every blood investigation performed irrespective of the needs. This results in an enormous wealth of data, often presented in the form of a computer print-out, and, I would say, very rarely looked at in the same professional manner as before. Recently, it was brought to my attention that a new anti-inflammatory drug, which had been approved for clinical trials, on which detailed protocols had indicated the age group of patient to be treated, and the exclusion of any patient with any abnormality in the blood, had been given to an 80 year old man who was already on treatment with Indomethacin and four other drugs and who had a platelet count of 80 000 cu/mm. After a few days he developed a haemorrhagic rash, and I was asked to advise on the cause of this. In the past, any haematologist carrying out a haemoglobin and film, and certainly if he was presented with a platelet/count of 80 000 cu/mm, would have repeated the tests and have carried out a clinical inspection of the patient. I believe the reason why this man's condition was overlooked was because hospitals are now generating so much information that to pick out from the mass of data what is important becomes almost an impossible task.

2. The second example I would like to give is the recognition of the nuclear sexing of polymorphs. (Davidson and Smith, 1954). It is surprising that in spite of the accurate quantitative determinations of every blood element, including neutrophils, carried out in every

hospital laboratory throughout the world, investigators had previously failed to recognize that a person can be sexed according to the appearance of the polymorphs. No quality control system as outlined in the FDA proposals would have ensured that such a distinction was made. It rests entirely on the quality of the observer to make such a discovery.

3. The third example I should like to mention highlights the difficulties in assessing whether a person is capable of doing the work according to their professional qualifications. Some 5 years ago we were investigating a new anti-inflammatory drug. During the dosing period clinical examination of the eyes in two species had not revealed any abnormality. A graduate scientist who examined the slides of the eyes of dosed animals noted a small change in the structure of the lens fibre. Some five pathologists looked at these slides blind and agreed with her conclusions. We showed the slides of the eye to an eminent ophthalmic pathologist at one of our leading University Hospitals who pronounced that the changes were not significant and that we could go forward with the drug.

We did not follow this advice: instead, we dosed animals for a longer period and examined them again by detailed clinical investigations and by histological examination. After approximately five months of dosing, typical cataracts were seen in the eyes of dosed animals. I believe that the essential element in the difficult recognition of this early change was the experience of the pathologist who examined the slides, and it enabled us to make a decision which, if it had been the wrong one, would have had profound consequences.

The quality control of the clinical chemistry carried out in Dr Fowler's Section, as I have outlined previously, emphasizes the importance not just of a result expressed in figures of a clinical chemical measurement, but also the considerable contribution of the sampling of the blood, the time of day at which it is taken, the housing of the animals, and all the other biological factors which can influence the result. To control these factors is entirely an internal matter—they cannot be covered by legislation—and require a persistent investigative approach to work.

I believe there is a real risk that the type of 'cookery book' approach to toxicology testing advocated by the FDA's Code of Good Laboratory Practice will not encourage further research into methodology and hence improvements in our technique. Instead, the inspectors who have been recruited and trained in toxicological procedures in an intense course problems which are inherent in biological experiments. At present the recommended number of animals per sex, per group, by the FDA includes 10 to 25 rats and 2 to 3 non-rodents. I find it very difficult to understand the logic in the differences of numbers of animals per group between rodent and non-rodent species. If we are only concerned with predictive safety testing, either we are using too few dogs or too many rodents, unless we are influenced entirely by the economics of testing. Likewise, I find it difficult to understand how a toxicity test carried out using Rhesus monkeys from the jungle where six per sex per group are advocated, can provide any meaningful information. To begin with, they are a risk to animal handlers because of B-virus, TB, and other viruses. Also it is impossible to age the animals accurately and to discover latent diseases. And why six per sex per group because we are using a sub-human primate? I believe it is important if we use a primate that we use monkeys which are bred for the purpose within the laboratory breeding complex. It is the only way we have of knowing the age of the animal and its disease status.

I also find it rather disturbing that an inspection team should, in 1977, pronounce on the wisdom of the methods used to test a drug in 1960. Methodology has advanced rapidly during that period and most responsible laboratories have changed their techniques according to the best available knowledge at the time. Let me give an example:

> It was not until October 1972 that the eminently useful document on carcinogenic studies of beta-blockers produced by Crout (Crout, J. R., 1972) became available and the importance of the level of the drug in the diet, its stability, its distribution, etc. were first recognized as important and required by the FDA to be determined in carcinogenic tests. I can find no reference to this fairly obvious quality control of diets recommended by the FDA or any other authority before 1972. Why this was not done is hard to say now, and even after 1972 the FDA have only recommended it for beta-blocking drugs. One omission from this important recommendation by Crout was that he did not request information regarding the absorption of the drug and it is only in the most recent regulation that we are required to show that the drug is absorbed.

I have raised these points to indicate that the science of toxicity testing is a gradually developing science and the science can only be practised according to the scientific information and recommendations which are available at the time of testing, and I believe that it is improper to judge tests carried out in 1960 by the standards which we lasting a few weeks will not begin really to understand the very difficult have reached today. It is also quite impossible to look into the mind of a scientist practising at that time to see what influenced his judgement as to whether a drug was safe.

In conclusion, I believe strongly that laboratories carrying out toxicity testing should take careful note of the recommendations made in the FDA document on quality control. I agree with most of them: there are omissions, and some technical points I do not agree with, and these I have tried to describe. However, I feel that there is little emphasis on the human element and far too great an emphasis on numeracy. If we are to advance the science of toxicology the important element is the quality of the professional scientists we manage to recruit and retain to work in this most challenging and important subject. Anything which is done which acts as a disincentive to recruitment of able scientists to work in this area would be a grave disservice to the safety of drugs in humans in the future.

References

1. Davidson, W. M. and Smith, D. R. (1954) *Br. Med. J.*, **2**, 6
2. Crout, J. R. (1972). FDA letter on requirements for animal carcinogenicity in all β-adrenergic blocking drugs. *Food and Drug Administration*

5

Quality control in an industrial laboratory

D. E. Stevenson

As an introduction the following points must be made:

1. The experience of my laboratory relates mainly to Agricultural Chemicals, Industrial Chemicals and Oil Products.

2. Quality control may be defined as the process by which line management ensures that research results are of the proper standard. Quality Assurance, on the other hand, is essentially an auditing process to assure that the quality of every aspect of an experiment is being controlled properly.

3. The FDA Good Laboratory Practice Regulations (GLP) are only one possible inflationary factor—new Guidelines from EPA and NCI will increase the number of animals per experiment. A wide range of percentage increases of manpower to meet GLP have been suggested—our first estimate of about 10–15% is below other predictions.

4. As the proposed FDA regulations are written, it seems doubtful whether non-US laboratories can comply, unless some discretion or reciprocal arrangements can be introduced, i.e. the regulations imply compliance with a range of other US laws.

5. The proposed FDA regulations will have a harder impact on industry than on some other types of institution—adverse results

may be acceptable whether or not obtained under GLP, 'proof of safety' must be by experimentation conducted under GLP.

6. European governments should study the FDA regulations in relation to the activities of their own industries because of the possible impact on both international trade and the provision of services.

Quality stems from an attitude towards science that goes far beyond the immediate practicalities of Good Laboratory Practice. Such regulations may increase the accuracy of documentation and enable one to prove, at a later date, precisely how something was done and by whom, but may have little beneficial effect on the actual quality of the work of the laboratory. A recent article in the *New Scientist* alleged that the manipulation of scientific information is not uncommon in academia. However, it is in toxicology, with the special connotation of health and safety that the public has been made aware of possible problems. During the last 4 or 5 years there has been an increasing debate about the quality of information upon which safety-for-use judgements are made. It is, of course, inevitable that as science advances, new knowledge and technology may make the interpretation of previous data questionable. What we are discussing, however, are situations where even by current standards, the approach is mediocre or unprofessional. No one is perfect— examples can be cited from the laboratories and scientists of governments, universities and public interest groups, as well as from industry. My own laboratory has found, by re-examination of original data, errors in published tables. Amendments were published, but these did not erase the embarrassment caused. This highlighted for us the need to ensure that all data are as nearly correct as possible.

Toxicological information is used for many purposes, not only for the avoidance of hazards, but also for political, commercial or personal reasons. The legal implications of information are also far-reaching and the potential for misuse only too frequent. In this type of climate, checks and balances are essential, not only for the protection of the public, but also for the protection of scientists themselves.

While the most publicized problems have arisen in the USA, other governments and international agencies have also drawn attention to less than perfect situations. An example of this is found in our own Pesticide Safety Precautions Scheme, (PSPS) where a relatively informal relationship exists between toxicologists representing government and those representing industry. We know that our official counterparts are anxious about the difficulty of interpreting some of the data submitted. Indeed, the British Agrochemicals Association recently circularized its members to stress the importance of following up effects noted in toxicological studies rather than merely to hope that a favourable official judgement would be made in the absence of data. Under the PSPS, an active dialogue is encouraged, but insufficient use is apparently made of this potentially constructive possibility. I have no doubt that this is not unique.

Quality relates to three interwoven aspects

(a) the individual scientist

(b) the organization that employs him

(c) the planning, execution and interpretation of experiments

All are important — inadequate science may be a symptom rather than the cause of the disease.

THE INDIVIDUAL

It is necessary to discuss how we perceive the proper approach of any professional scientist working in a toxicology laboratory or who may be called upon to review data obtained by other scientists. Many professions have a code of conduct that outlines the ethical factors to be weighed. Every toxicologist is a potential expert witness in a court of law or in a situation such as a public hearing in the USA. His standing will depend in part both on the quality of his science as judged by the scientific community and on the quality of his decisions. The individual must decide where his prime duty lies when there is a clash between professional opinion and other demands. It is easy to say the former must be pre-eminent, but unless the organizational structure takes potential conflicts into account, opinions may get attenuated or completely stifled before they reach the appropriate management level. The pressures on scientists who enter in public debates are such that many reasonable people are reluctant to participate since they feel that their independence and judgement may be wrongly compromised.

When dealing with scientific issues, the individual must consider whether the subject falls within his knowledge and experience and also whether the experimental evidence has a sufficiently sound scientific basis to permit a well-founded opinion to be expressed.

Thus, in quality control, we are not only concerned with the accuracy of data and the scientific merit of the experimentation, we are even more concerned about the quality of interpretation and the way research information is utilized and disseminated. In this context, the generally accepted control of quality in science is publication in the scientific literature. It is unfortunate that a considerable amount of toxicological information is not subject to this process. It is equally unfortunate that many of the analyses that cast doubt on existing data remain essentially unchallenged because they also have a twilight existence.

THE ORGANIZATION

Both industry and governments engage in toxicological research and the interpretation of results is part of their activities. It is self-evident that in both sectors first class information and advice is essential. In our view these relate to two aspects, namely the definition of the inherent toxicological properties of compounds and the interpretation and application of this information in the practical situation. For the former, a detailed current knowledge of laboratory work is essential, which is best acquired by being actively engaged in research. For the latter, a knowledge of the practical use situations is needed. This is best acquired by being more closely associated with the 'market place' than is feasible for laboratory scientists.

Companies state their corporate objectives and seek to obtain a satisfactory balance between them. Such objectives are often stated in terms of profitability, social responsibility, obligations to employees and continuity. Each of these areas has a direct relationship with the quality of toxicology, which has to be assessed also in the light of national and international attitudes towards social and economic standards, legal requirements and the quality, type and scope of available scientific resources. Quality standards, therefore, may not be uniform, but should be the best possible in a given situation.

Toxicological information may be required for two major reasons

(a) Product registration, manufacturing permits, etc., i.e. regulatory reasons

(b) To assess what recommendations for the protection of health are necessary.

This distinction may appear artificial and a matter of degree, but in practice may have a major impact on the amount and quality of the studies carried out. All types of research are expensive, not the least toxicology, and companies do not wish to spend funds unnecessarily. The short sighted policy that may be adopted is to obtain only the minimum data and to shop around to find which laboratory is willing to provide marginally adequate information, producing the highest no-effect levels at the cheapest price. A comparable approach has also been adopted by some regulatory agencies and public interest groups, only in this case it has been manifested by commissioning experiments with the deliberate intention of obtaining qualitative adverse information without attempting to quantify safe concentrations. There are examples of such data being kept 'under wraps' where the results were not adverse!

It is not easy for a toxicologist to exercise proper judgement where his role may simply be to ensure that a tick can be placed in each box of a check list of information requirements. Nelson's blind eye won his battle not because he chose not to see the enemy ships but because he chose not to see the signal which instructed him to stop—perhaps a lesson for us as well.

THE CONDUCT OF RESEARCH—THE LABORATORY

Toxicology is not an armchair science. We believe it is the responsibility of a laboratory to translate the requirements of a development project or the solution of a toxicological problem into a research programme, which can then be discussed and agreed with the business centres and their professional advisers. The responsibility for the quality of the experimentation and the problem solving capability of the laboratory lies firmly with the laboratory management. However, the interpretation of the experimental information into the practical context of the business and the subsequent advice on hazard avoidance may be more appropriately the responsibility of those nearer the use situations. In the case of the Royal Dutch Shell Group, this is one of the functions of the Toxicology Division in the Hague and London. Thus, the description of the toxicological properties of a compound and the assessment of hazard are vested in two separate organizational groups. This minimizes the temptation to take the attitude that because an observation may not suggest an immediate practical hazard, it should not be documented and, therefore, save the business some embarrassment. A laboratory will only contribute to the full when the business has full confidence in its sense of responsibility and ability to conduct work to the highest standard. This is a very sensitive relationship which does not just happen, it requires active management from both sides.

One of the fascinations of science is the continuing learning process. A laboratory can be a collection of separate individual endeavours, or it can be a more tangible structure, where the total ability and experience is greater than a mere summation of the individuals. Toxicology with its links with many disciplines is particularly susceptible to changing patterns of teamwork, embodying both continuity and the most appropriate expertise. We have arrived at a managment system which depends on the line organization of resources along broad scientific disciplines—experimental toxicology, chemical toxicology and pathology, with a matrix organization of project teams chosen by potential contribution. These teams (not committees) have access to all the relevant information and there is active participation by toxicologists, chemists and pathologists at this level in determining the recommended experimental approach for review by the laboratory management, i.e. the experimental approach flows upwards and is not imposed from the top downwards.

The FDA Regulations on Good Laboratory Practice have arrived at a timely stage in our thinking on quality control. Until this point we have regarded this as part of the function of the individual sections—with real success. We are now considering a third dimension in our management: the development of a system that monitors the quality of the laboratory as a whole, i.e. both line resources and project teams. This implies a much more comprehensive approach than is required by the GLP itself. We believe we could fulfil the immediate principles of GLP relatively quickly and with a minimal fall in productivity. However, the discipline of studying the regulations has crystallized the more diffuse views which are already emerging.

The application of the FDA regulations to non-US laboratories is still a matter for discussion, particularly where existing practices or regulations are in potential conflict. We are assuming, however, that we should at least match the spirit of the regulations.

About 6 months ago, we requested a senior colleague to review the circumstances which led to the FDA proposals and to discuss the implications with the appropriate scientists within the laboratory and to make recommendations to its management. This review has now been completed. The clear message that has emerged is that more attention must be paid to the co-ordination of data storage and the documentation of quality control, e.g. in most cases the right information is available but may be held informally and is therefore liable to be lost when staff changes occur.

We are now

1.1. Studying the policy decisions that may be required in relation to coverage, etc.

1.2. Identifying and defining those activities that will be included.

1.3. Consulting with the staff on the implications of GLP with the aim of obtaining their commitment.

Three areas of action are expected

2.1. Standard operating procedures can be modified or drafted for defined activities.

2.2. A skeleton Quality Assurance System will be developed by grafting on to existing activities, particularly in the data handling area.

2.3. A project team of senior staff will be established to draw up system capability requirements and to oversee the subsequent steps.

We are planning for (2.1.) and (2.2.) to be completed within 6 months, but expect that a minimum of two years may be required for the systems capability study and implementation to be completed. I suspect this may be optimistic.

Finally, I would like to outline our present approach to quality control in some aspects of long term studies, since this is the area which has received the greatest criticism in the USA. Disaster may be more instructive than success. Our own attitudes have been modified very substantially by our experience in the 1974 EPA public hearings on aldrin and dieldrin. It must be clearly recognized that in such situations the slightest room for criticism will be seized upon and magnified, since some mud will stick, no matter how unwarranted the initial allegation. Good Laboratory Practice may change the battlefield, it will not resolve the conflict.

SUPPLY OF EXPERIMENTAL ANIMALS

Within months of opening our Laboratory in 1960 we found that the purchase of 'SPF' animals from a commercial supplier, and then breeding and dosing them under minimum barrier conditions, was not a satisfactory situation—animals were dying of respiratory infections even in 90-day studies. We made a decision to set up our own breeding unit and develop a much stricter control of long-term animal facilities. These are described by Walker and Poppleton[1] and by Walker and Stevenson[2]. The disease status of the breeding colony is monitored by regular bacteriological and parasitological examinations. In addition, many of the animals on experiment, and some of the discarded breeding animals, are subject to pathological examination. While it is a fact that each rat and mouse bred in our own facility costs marginally more than if we purchased from outside sources, the advantages in terms of health of animals, knowledge of ancestry and the inbuilt safeguards of in-house stock control, make the small cost difference fade into insignificance. Our control of infectious disease is by a slaughter policy and re-stocking with new healthy stock. In order to minimize disruption to programmes, we have three separate long-term rodent units and are building a second rodent breeding unit. The production of animals is under the supervision of a veterinarian, who is also responsible for monitoring the quality of the animals produced.

We remain to be convinced that there is any other satisfactory mechanism to obtain animals with identified characteristics and qualities. The use of litter mates in long-term experiments can decrease significantly the variation between groups and increase the sensitivity of the experiments to detect changes.

STAFF AND ROUTINE OBSERVATIONS OF LABORATORY ANIMALS

The ultimate quality of long-term studies depends heavily on the quality of the animal technicians and on the way animals are studied during the course of experiments. Trained and experienced staff are essential and this implies continuity of employment. The observations made by the animal technician assigned to an experiment are checked by the technical supervisor of the unit, and also by another senior animal technician or graduate at least once a weeek. We have always housed animals in single cages (a) to prevent cannibalism; (b) to allow more satisfactory observation; and (c) because the number of animals per cage affects food intake and body weight. We are aware that there are some valid scientific arguments against caging rats singly. However, it is our conscious belief that the advantages outweigh the disadvantages in what must be an imperfect situation. All animals are observed daily and the regular weighing procedures are an opportunity to study the condition and behaviour of animals in more detail. An experienced animal technician may be very expert in detecting very slight changes—there must be many anecdotes on this point. It would be a great mistake if the graduate supervisor did not also regularly examine the animals and assure himself that the experiment was being conducted properly. This can only be accomplished by actually looking at and handling animals!

We have now adopted a pathology form that is prepared at the beginning of each experiment and is retained in the animal unit. Any abnormality noted on an animal is entered on a sheet of gummed tear-off slips which are subsequently transferred to the pathology form, so that a complete clinical record is available with the animal, particularly when autopsied.

PATHOLOGY

I am prejudiced against any laboratory that does not have at least one experienced pathologist. Moreover, I believe that he or she must be an active member of the project team at all stages. Even though autopsies may be carried out by technical teams, he must review their findings and determine what additional observations or material are needed. A satisfactory final diagnosis can only be determined by an evaluation of all the facts, clinical, gross pathological and histological.

Quality relates to (a) the standard of autopsy, which can easily degenerate into the rapid dismemberment of a carcase and the weighing of organs; (b) the sampling of tissues for processing; (c) the preparation of routine histological material—not uncommonly poorly fixed, badly cut and patchily stained; (d) histological examination of all animals; (e) follow-up studies. Some aspects of Quality Control are obvious, e.g. the quality of slides can be ascertained easily. The crucial area is perhaps the control of the quality of diagnosis. We ensure that at least when the pathology on a study is completed, the data—including slides— are reviewed by the Head of the Division and abnormalities may be debated on an even wider basis, and if necessary additional follow-up studies instituted. This avoids the embarrassment of a subsequent analysis by another pathologist, perhaps also on the staff of the same laboratory, arriving at a different set of diagnoses.

Another pet aversion is the current fashion of flying a few selected slides around the world so that consultant pathologists can make their own interpretation of the results, usually without recourse to any of the detailed information of the clinical history or the autopsy findings of either the treated or control animals.

We retain fixed wet tissues, paraffin blocks and sections on a permanent basis, although the wet tissues are destroyed when they deteriorate too far.

HAEMATOLOGY AND CLINICAL CHEMISTRY

These investigations are highly automated and it is possible to produce reams of reproducible rubbish unless quality control procedures are included. These are of two types: (a) the assurance that the methods developed for human blood are equally applicable to animals—by no means always the situation; (b) the use of known standards at frequent intervals and, where possible, participation in inter-laboratory quality control schemes. Unfortunately, there are laboratories who do not observe these simple procedures.

DATA STORAGE

During the preparation of statements for the 1974 aldrin/dieldrin hearing, we came across some errors in tables that we had published. In order to prevent a recurrence of this, we made a specific individual responsible for checking and storing centrally all long-term animal numerical data and records. Thus data are now checked at least twice, once in the originating unit and again in the data storage unit before the data are checked and analyzed by the statisticians. Some countries already request routinely all the individual information and we are in a position to provide printouts of all the numerical data on long-term experiments. (A few early experiments are being re-examined and the presentation of data up-dated, with suitable checks.)

PREPARATION OF REPORTS

The preparation of reports against a tight time schedule, with a limited time for scrutiny and thought, leads inevitably to errors. There may also be a psychological problem relating to the production of unpublished reports on single experiments compared with the preparation of a paper covering a research programme and which is subject to an external editorial process. Scientific papers abound in comments which allow room for a variety of interpretations, and also a range of additional experiments which would be needed to provide a definite answer (which, like the end of a rainbow, is seldom reached). The reporting of toxicological experiments requires a more exact approach. One particular problem area is the observation of statistically significant differences which may well be due to chance and where the interpretation is an overall matter of judgement.

This list of aspects is by no means comprehensive, it is given to provide an indication of some of the areas of quality to which we have given considerable thought. We have found that good scientists pay attention to the quality of their work. However, in multidisciplinary projects such as long-term studies, it is important that a mechanism is provided to ensure that each input is adequate, because it is at the weakest link that the chain will break.

References

1. Walker, A. I. T. and Poppleton, W. R. A. (1967). The establishment of a specific pathogen-free (SPF) rat and mouse breeding unit. *Lab. Animals*, **1**, 1
2. Walker, A. I. T. and Stevenson, D. E. (1967). The cost of building and running laboratory animal units. *Lab. Animals*, **1**, 105

6

Good Laboratory Practices (GLPs) and the Bioresearch Monitoring Program

William M. Merino

In order to provide some perspective for the basis and need for Good Laboratory Practices (GLP) regulations, I would first like to provide you with background information on Searle's experience that led to the development of a Searle draft of Good Laboratory Practices in January, 1976 and FDA's proposed regulations published on November 19, 1976.

In July, 1975, during a hearing chaired by Senator Kennedy, certain FDA employees made allegations of improprieties on the part of Searle in conducting and reporting animal safety studies to the Food and Drug Administration. In order to resolve these allegations of improprieties, FDA undertook a thorough investigation of preclinical research at Searle. Searle agreed to co-operate fully with such an investigation.

The scope of the investigation included identification of selected animal studies conducted by or for Searle which were submitted to the FDA after January, 1968. The criteria for selecting products were based on the number of people receiving the drug, and the length of recommended treatment.

Immediately following the July, 1975 hearing, in order to facilitate FDA's review of the massive documentation supporting the animal reports that had been submitted to them, Searle called on its information scientists to organize in a central location all raw data supporting our preclinical safety studies. Prior to this time, our documentation was located throughout the research facility in departmental files or the files of individuals who had generated the data. We thought it desirable to locate all documents under one central control, so that to the extent possible, FDA requests could be answered with assurance that all requested documents were being produced for FDA review, and in order to respond quickly and accurately to the anticipated FDA requests. We utilized a relatively small group of people with the ability to organize properly a massive number of documents for ready access and retrieval. These individuals made a thorough search of all files throughout the research facility for documents relating to Aldactone and Flagyl. As the task proceeded, it became apparent that to complete the compilation of documents, additional personnel would be required. Consequently, the group was expanded to over 300 individuals from various departments, who assisted in collecting and organizing documents relating to Searle drug products. This task was essentially completed before the investigation began.

On October 6, 1975, the investigation at Searle began. It was concluded on December 9, 1975. Concurrently, an investigation similar in scope began at Hazelton Laboratories, a contract laboratory Searle used to conduct animal safety studies. At times, there were more than 20 FDA employees at Searle during this massive and unprecedented investigation. The FDA employees were divided into teams with each team responsible for reviewing different studies. Each team was composed of an FDA reviewing pharmacologist and two (or three) Good Manufacturing Practice (GMP) field investigators whose experience was related to the review of process-orientated records. Prior to the on-site visit, the investigators were given a 'crash' course at the National Center for Toxicological Research (NCTR) to obtain FDA's view of what it considered to be good laboratory practices. NCTR, which is part of the FDA, is considered by FDA as a 'model laboratory'. Searle assigned two top level scientists to work with each team sent by the FDA. These Searle 'escorts' aided the investigational team on a daily basis providing documents for them, setting up interviews, and explaining the systems and processes in use at Searle. My responsibilities included the direction of our scientists in providing the FDA with needed information.

Each FDA team began its investigation with a tour of Searle's research facilities and a request for documents supporting particular animal studies, final reports of which had been previously submitted to the FDA. Naturally, each of the FDA teams expected its request for documents to be fulfilled promptly and efficiently. In this regard, the prior organization of all documents in a central area became extremely important. Given the nature of the requests, many of which involved unreasonable short-time demands, it was unlikely that the FDA requests could have been fulfilled without prior organization of documents for ready retrieval. A further value of the central location was the ability of the information scientists to keep track of the requests, who had what documents, and to check that all documents were returned. In order to indicate the massive undertaking of organizing these documents, it is important to realize that in some cases, FDA was looking at preclinical safety studies that had been initiated over 10 years previously. Although the central location had been organized to facilitate the FDA, the FDA inspectors were reluctant to rely completely on the fact that all documents had been collected for storage in this area. Consequently, to assure that they had received all documents relating to a study, some teams personally inspected employee's files and desks, looking for possible documents that were not in the central area.

In addition to the review of documents, FDA conducted over 200 interviews of Searle employees in order to resolve questions on practice and procedure at a particular point in time. During interviews, employees, because of the unpredictable nature of the questions, were without benefit of immediate access to prior documentation. They, nonetheless, were asked to recall precise facts relating to specific animals. In some cases, details were asked on facts that occurred 6 or 7 years previously. As one might imagine, given the lapse of time, recall, if there was any, was not instantaneous. Numerous subsequent referrals to documents were necessary to respond to the questions.

A total of 25 animal safety studies were evaluated. These included chronic and reproductive studies on the following products: Aldactone, aspartame, Norpace, Flagyl, Cu-7, Ovulen, and a component of Syncro-Mate B.

The FDA review of documents and interviews covered literally every aspect of animal research. The qualifications, training and supervision of all employees from animal caretaker to pathologist were questioned. Quality control data relating to the compounds being tested were requested. Data relating to the testing of animal feed were requested. Records relating to the purchase and care of animals were analysed. The investigators questioned the technique of feeding, identification, and care of animals. Records relating to observation of animals, feeding, and body weight were closely scrutinized for any inconsistencies. All records were checked for the signature of the author and the date of observation. Procedures within the Histopathology Laboratory were questioned.

Statistical calculations were checked against raw data. Reports were checked for internal inconsistencies and compared with prior drafts of the reports. Equipment and facilities, including such things as light, space, temperature and ventilation were checked. Gross and microscopic pathology reports were checked for inconsistencies and for any deviation from raw data. Grammatical or editorial deviations were noted. Review procedures relating to submission of reports were questioned.

In late October, when Searle analysed the investigators' activities, it became clear that: (1) Laboratory activities were being viewed as if they were manufacturing activities in terms of record keeping and documentation. (2) Many so-called 'violations' cited were based on the opinions of the FDA pharmacologists about how toxicology studies should ideally be performed for design, execution, and record keeping in the absence of regulatory provisions and universally agreed-upon standards. Whereas regulations had been promulgated for good manufacturing practices, none had been provided for laboratory practices. The FDA's criteria were based on standards of regulation in different areas of activity in which laboratories operated on the basis of diverse scientific expertise.

A group of six people from Searle—five scientists and one lawyer—were assigned to tasks of developing a document entitled 'Good Laboratory Practices' which was expected to deal with documentation of laboratory activities and the provision of standards against which scientific activity could be evaluated. This group of individuals worked for approximately two months developing the initial draft which was submitted to the Food and Drug Administration and the Pharmaceutical Manufacturers Association in January, 1976.

At subsequent hearings by Senator Kennedy in April, 1976, Commissioner Schmidt requested additional funds and employees to review and monitor the reports of preclinical testing laboratories. Commissioner Schmidt suggested that all pharmaceutical companies and laboratories might be inspected. As a result, the US Congress approved 16.6 million dollars and 606 new positions specifically for monitoring bioresearch. This programme has become known within the Agency as the Bioresearch Monitoring Program.

In August, 1976, the FDA released for comment by specific governmental agencies draft GLPs similar in many respects to the Searle draft. Searle was advised by Agency personnel that its draft had been used by FDA in preparing the GLP proposal. On November 19, 1976, proposed GLP regulations were published in the Federal Register for comment on or before March 21, 1977.

Prior to final comments on the proposed regulation, a public hearing will be held on February 15, 1977. The purpose of the hearing is (1) 'To provide an open forum for the presentation of views concerning the merit of the proposed regulations and their general applicability and practicability and (2) To foster greater consideration of the proposal among the scientific community, the regulated industry, and the public.'

In the interim, the FDA has instituted a GLP pilot programme which began in early December, 1976. This pilot programme is designed to:

1. 'demonstrate to the industry that the FDA intends to implement a programme of non-clinical laboratory inspections.'

2. 'develop baseline data to measure the conformance of non-clinical laboratories with the *proposed* Good Laboratory Practice Regulations.'

3. 'obtain additional knowledge about the current status of laboratory practice to improve the quality of the proposed Good Laboratory Practice Regulations.'

4. 'take appropriate action whenever a situation involving a serious violation of the proposed GLP Regulations is encountered or when fraud or other deliberate falsification of test data has occurred.'

The programme provides for the inspection of 42 non-clinical laboratories. Detailed forms and reports are to be completed by the inspection teams. The results of these inspections will give the FDA a better perspective in their assessment of the GLP regulations. However, it must be recognized that these inspections are not contemplated to be as intense as the inspection of Searle; nor could they be, considering the differences in manpower, time and circumstances.

Regardless of what happens during this pilot programme, or what comments are received, I believe Good Laboratory Practice Regulations will become law in the United States sometime after March 21, 1977. Later, I will discuss what I think this will mean to those who conduct studies to support applications for research or marketing permits for products regulated by the US Food and Drug Administration.

WHAT ARE GLP REGULATIONS?

The FDA states that GLP regulations are 'methods to be used in, the facilities and controls to be used for, conducting nonclinical laboratory studies to assure the quality and integrity of data filed which become the basis for regulatory decisions made by the Agency.'

The principal benefit of the GLP regulations is a statement, previously not available, of what is expected by the FDA from the investigating laboratory. In part, the GLP regulations establish procedures for record keeping. They are designed to look for deviations from the routine. The GLP regulations are 'process controls' to assure that the end product, that is study results, are controlled. There is a strong analogy to Good Manufacturing Practice Regulations whereby the processes used to reach the finished product are strictly controlled. The intent of FDA's GLP programme is apparently to monitor the monitoring of the process, and it is not their intent to monitor the process directly. The FDA will, however, also continue to conduct independent data audit inspections which will directly monitor the end product.

The FDA states that the GLP regulations will not deal with the interpretation of the study or study results. This may not be an easy task based on Searle's past experience with the FDA investigators who had difficulty in separating their personal opinions from the function of reviewing the process control.

Searle is now developing standard operating procedures, modifying facilities, equipment, personnel and other procedures so as to be prepared for the final Good Laboratory Practice regulations. As I stated before, Searle is, in principle, in favour of GLP regulations. They will provide a firm foundation of what is expected from the laboratory and what the Agency is entitled to demand. They will also apply objective standards of laboratory performance uniformly and equitably to every testing facility.

It is Searle's position that GLP regulations combined with clarification of existing reporting regulations will yield at least 5 major benefits to those conducting preclinical research:

1. Accurate knowledge of what is expected of them in the form of records both as to nature and content.

2. Knowledge as to techniques to be used in making entries in laboratory records.

3. Ability to respond to requests by the Agency promptly in a format familiar to FDA personnel.

4. Confidence in the availability of the records necessary for future analysis of results and Agency review, while at the same time per-

mitting maintenance of a realistic and reasonable record retention system.

5. Simplification of management control or directives to personnel engaged in preclinical studies.

I do not plan to describe the details of what Searle is doing nor am I going to tell you how to develop standard operating procedures. Procedures will vary from lab to lab and each facility should develop its own operating procedures in order to have a system that works optimally under its own circumstances. It is also not my intent to review and discuss the proposed GLP regulations in depth. I am assuming that all of you have read and are familiar with the proposed regulations. However, at this point I would like to discuss with you those aspects of the proposed GLP regulations with which Searle disagrees and the changes which we plan to propose.

1. Quality assurance unit

The proposed regulations state the following:

'(a) A testing facility shall have a quality assurance unit composed of one or more individuals who shall be responsible for assuring that the facilities, equipment, personnel (including personnel safety), methods, practices, records, and controls are in conformance with the regulations of this part and for assuring the quality and integrity of the data obtained from a non-clinical laboratory study and for adherence to protocols and standard operating procedures. Each facility's quality assurance unit shall also monitor the quality and integrity of any non-clinical laboratory studies or portions thereof done by contractors or grantees.'

The regulations go on to describe in detail what the QAU shall do. This can be summarized as follows:

(a) Maintain a copy of a master schedule sheet, protocols, and standard operating procedures.

(b) Inspect each phase of the study periodically and maintain written reports of such inspections.

(c) Perform a complete evaluation every 3 months of all phases of all studies lasting more than 6 months. Studies lasting less than 6 months shall be evaluated more frequently than every 3 months.

(d) Submit periodic reports to management and the study director.

(e) Assure there are no deviations from protocol and standard operating procedures.

(f) Review the final study report.

The proposed regulations also state that all records maintained by the QAU shall be made available to the FDA.

One can readily see that the FDA's concept of the function and responsibilities of a QAU are quite detailed and inflexible. What is not totally clear is whether or not a QAU is the best means of assuring that such activities are done. Searle believes that a QAU as described in the proposed regulations, should not be required. The function of the QAU (as defined by FDA) would remove, in part, the responsibility of data reliability from the scientist and give it to the QAU. Depending on the qualifications and expertise of those people comprising the QAU, which by the way, are not defined by the FDA, there is a danger that these individuals could inadvertently create an obstacle to the professional judgement constantly required during *in vivo* and *in vitro* testing.

It is Searle's position that GLP regulations only require that the management of a testing facility develop a system based on the personnel and needs of the facility that will assure compliance with GLP regulations. Mechanics of how that is done should reflect individual management style and in so doing, the entire assurance system becomes more effective.

We are also concerned that internal audit reports should not be available to FDA investigators. A requirement that they be available will result in extreme reluctance to include comments which could be interpreted against interest, thus greatly reducing the usefulness of such audit reports.

The sponsor's quality assurance procedure should not be required to inspect every phase of every study, but should assure GLP adherence through *periodic* inspections of all phases of laboratory practices.

The proposed regulations also state that 'the QAU shall also monitor the quality and integrity of any non-clinical laboratory studies or portions thereof done by contractors or grantees.' If *every* facility is required to have a QAU or some other management system, then it is an unnecessary duplication of time and effort. The responsibility of the sponsor should only be to assure that the facility has an effective QAU or other appropriate management system.

2. Study director

The proposed regulations state the following:

> '(a) Before the initiation of a non-clinical laboratory study, a scientist or other professional of appropriate education, training, and experience, or combination thereof, shall be identified as the study director. The study director has ultimate responsibility for implementation of the protocol and conduct of the study . . .'

The FDA places the responsibility of the study director to one person. Many situations exist where scientists from more than one discipline can have a major responsibility in a study. For example, the protocol may include pharmacological and/or pharmacokinetic evaluations. Therefore, we believe that there can be more than one study director and suggest that the responsibilities of each be outlined in the protocol.

3. Mixtures of substances with carriers

The proposed regulations require that each batch of test or control substance with a carrier mixture must be assessed for stability characteristics.

We believe that the stability of *every* batch/carrier mixture should not be required. If the stability of the test or control substance has been established at one time, for example, in saline or water, it is not necessary to repeat this for every batch.

4. Animal care

Searle believes that the rigid application of the Animal Welfare Act with regard to the housing of animals is unrealistic. This would require many facilities, especially those in non-US laboratories, for example, to purchase new cages resulting in a substantial cost burden. For example, how will academic laboratories, such as those at colleges and universities be limited by these standards for facilities. The calibre of the professional may be extremely high, but laboratory and animal facilities may not conform in every respect to the regulations.

Since I am, so to speak, a foreigner among you, you may be asking yourselves, 'What do these new regulations mean to me?' Those of you from contract laboratories are probably more aware of the implication of the GLP regulations. Many non-US contract laboratories conduct animal safety studies to support US research and marketing applications. Others here may work for subsidiary companies whose US parent companies rely heavily on your studies for support of US INDs, NDAs, NADAs, Food Additive Petitions, etc. More importantly, some non-US companies may file applications directly to the FDA.

Clearly in these cases where non-US studies will be used in the US, GLP regulations will probably be applicable. Otherwise, testing of products for FDA clearance will be driven to laboratories outside the US.

Dr William D'Aguanno, an FDA official who has been instrumental in the development of GLPs, has given a talk to the Association of the British Pharmaceutical Industry. It is my understanding that his talk was well received and those who heard him seemed to agree in principle to the idea of Good Laboratory Practice Regulations.

To date, officials from the FDA have met with the CSM and Canada's Health Protection Branch to discuss co-operative efforts in the development of GLP regulations. I do not know what agreements have been reached as a result of these inter-agency discussions. Perhaps some of you are more knowledgeable of these discussions.

In December, I spoke with one FDA representative who said it was their hope that the GLP regulations, when finalized, could be applied to non-US testing facilities which conducted studies for support of US INDs and NDAs. However, like other US FDA regulations, they cannot be enforced in overseas countries, especially if they conflict with local regulations. The FDA's alternative is to enforce them via the US sponsor, by withholding marketing approval, rejecting studies, or removing already marketed products.

At this time, it is my understanding that the FDA is not proposing any routine inspections for non-US laboratories. However, they do plan to conduct data validation inspections of animal safety studies probably prior to either NDA approval, other application approval, or during Phase III studies. If during such an audit, the FDA inspectors find any deviations from GLP regulations or serious data discrepancies, they would be identified to the laboratory and the US sponsor. Significant deviations may result in action by the regulatory agency.

Some non-US countries do not now permit GMP inspections for those products intended for US sale and manufactured outside the US. In these situations, their own regulatory authorities inspect the facility and provide a report to the FDA. A similar situation may be applicable and satisfactory with respect to GLP regulations. However, it is FDA's intention to discuss the scope of such inspections with other drug regulatory agencies.

As from December, no non-US laboratories have been inspected by the FDA; and therefore, the quality assurance aspect of studies cannot be determined. I have heard that representatives of one British contract laboratory say they would love to have someone from the FDA inspect their facilities. The FDA feels that most overseas contract laboratories will have the same attitude for obvious economic reasons.

Overseas facilities would most likely receive advance notice of GLP inspections from the FDA when they do occur. Both the facility and US sponsor would be notified prior to any visit.

What should you do to prepare for a visit from FDA inspectors who will be looking for deviations from Good Laboratory Practices? First, you should obviously be familiar with the GLPs. I would advise you to have your legal counsel review these and be present at any such inspections. You should also become familiar with the FDA's Good Laboratory Pilot Program which is currently under way in the US. This should give you some insight as to what to expect if and when you are inspected to determine conformance with GLPs. I hope a copy of this compliance programme can be made available to you.

For the remainder of my talk, I would like to highlight portions of the Searle investigation and the proposed GLP regulations and ways you can prepare your laboratory for possible inspection.

1. Record Retention

As I mentioned at the beginning, Searle spent a great deal of time and resources gathering all records and data on each study into one central area. Over a period of 10 years, many of the records were distributed throughout the company as a result of people leaving, area relocation, and reorganizations. You should take the time to see if study data and information can be easily found in your facility. It would be advantageous to centrally locate these documents, check for completeness, and index them for easy access.

These records should include all raw data, protocols, specimens, final reports, and other required documents pertinent to the conduct of the study.

You should have written procedures to document the flow of data from the laboratory through analysis to the final report. It should be explained who does what, how it is checked, and how the system is monitored. If you do not have written procedures for this, sit down and try to explain verbally to someone how it is done. You may find this difficult to do in the absence of written procedures.

2. Documentation

After you centralize and index all your data and records, check to see that all information is available. If pieces are missing, why are they? Can you explain? Review the raw data sheets. Are numbers or values missing? Are they crossed out? If a value to be recorded is missing, there should be an explanation as to why. Any crossed out or marked over data with new values inserted must be explained. It should be initialled, dated, and reasons for the change explained.

If there is no documentation, perhaps you feel that there is a good explanation. However, during the Searle investigation, the FDA remained dubious even though explanations were presented.

3. Protocols

A protocol must be available for each animal safety study conducted. The protocol must be detailed and not just an outline. Most importantly, the protocol must be adhered to. If there are any changes, the protocol should be amended appropriately in writing, documenting the reason for the changes, who authorized them, and when. If the study was done at a contract laboratory, the nature of the input by the sponsor should be fully documented.

4. Test materials and quality control

The purity of each test agent should be determined prior to use in the study. Impurities should be quantified. Records should be available.

The stability of each agent should be determined under the same conditions as the agent will be administered. Records should be available.

The homogeneity of test substance with the carrier, for example, feed mixtures, should be determined from freshly mixed batches during the course of the study. The mixing and formulation procedures should be adequate and records should be available.

5. Personnel

Curricula vitae should be available for supervisory, professional and technical personnel involved in each study. Job descriptions and qualifications of personnel at the time the sudy was done should be on file. Organization charts including responsibilities of each individual should be available.

Training programmes including educational courses and on-the-job training offered to laboratory personnel should be described. An effort should be made to retain continuity of all personnel throughout a study. The reasons for changes should be documented.

Make sure that every level of personnel involved in a study from those who sweep the floor to technicians and supervisory personnel understand the purpose of the research and know how minor deviations from protocols or record keeping can devastate a product.

SUMMARY

I have given you a brief history of the development of GLP regulations and an overview of what can be expected if you are inspected. Clearly, those of you conducting animal safety studies for support of US INDs and NDAs can look forward to close review of all reports submitted to the FDA and possible GLP site inspections.

US industry has learned to live with governmental regulation, and I am sure will continue to do so. Several years ago at a scientific meeting, I can remember Dr Crout, Director of the FDA's Bureau of Drugs, stating that he likes regulations because they make it easier for the Agency to regulate industry across the board in order to obtain uniform standards.

Many non-US regulatory agencies are beginning to follow the US pattern of stricter regulation although they may be reluctant to do so.

In addition to GLP regulations, the Bioresearch Monitoring Program has two other aspects that will result in even more regulation and will have a substantial impact on *clinical* studies conducted outside the US. The first part of this involves the monitoring of clinical investigators, sponsors, and monitors. This programme will be analogous to the monitoring of animal safety studies. Standards and regulations will be developed, clinical investigators and industry will be audited as part of a compliance programme that will be initiated soon.

The second part will be the review of function and process of institutional review committees. An inspection programme will begin soon.

In the interim, prior to finalization of all these regulations, enforcement will continue to be based on data audits in the area of animal safety and clinical investigation. The major outcome will probably not be prosecution, but rather non-acceptance of studies.

SECTION THREE

The Independent Expert
Chairman:
J. McL Philp

7

Quality and integrity assurance
J. Jacob

The French regulations for quality control have existed for nearly 20 years, when it was decreed that the non-clinical laboratory studies would be considered in support of an application only if they had been performed by 'agreed experts'. These regulations apply to proprietary medicinal products.

In contrast to the rules proposed by the FDA, this regulation was and is still essentially person-orientated and not process-orientated. The experts are proposed to the Minister of Health by a committee. The modalities of these proposals, the duration of the mandate of the expert etc. have varied. The last ones have been given in the ordinance of November 21st 1972 which stipulates 'the committee must ascertain that the candidates for the office of expert present the necessary guarantees of honourability, competence, and good reputation in their respective disciplines and that they possess appropriate academic qualifications'.

This 'person-orientated' character is strengthened by the following stipulation of the ordinance 'the experts cannot have any financial interest, either direct or indirect, even by interposed persons, in the marketing (commercialization) of the drugs they investigate. They also are not allowed to perform this service for the manufacturer of whom they are salaried employees'. They are nominated now for 5 years (before it was

3 years) but can be dismissed by the Minister of Health on the proposal of the committee, the members of which cannot be experts themselves.

Thus, the independence of an expert is compulsory and a system of qualification and disqualification is already in force, and has been since 1960. It is to be stressed that the French experts need not legally be French citizens. Indeed, experts of other nationalities have been nominated especially since the Council of the European Economic Community has given 'directives' to bring together the legal control of the pharmaceutical 'specialities'. Although 'person-orientated' the French system is secondarily process-orientated as the committee 'must ascertain that the candidate has at his disposal working facilities sufficient to carry out the study in conformity with general rules provided'.

These general rules referred in particular to compliance to protocols and very recently, in September 1976, the regulation has placed more emphasis on protocol as the tests of the directives of the Council of the European Communities have been promulgated in a French ordinance. This ordinance no longer speaks of an 'agreed expert' but of an 'experienced investigator'; nevertheless, it is not specified that the former regulations concerning the experts are abolished. Indeed these protocols present some of the rules proposed by the Commissioner of Food and Drugs, but in a much more general way and without any sanction other than the refusal of a visa to the proprietary medicinal drug; the qualification of expert can also, in principle at least, be lost according to what has been said above.

Three items of this ordinance merit some comments: the characterization of the test substance, the establishment of the protocol and the report. The French regulation has recognized a long time ago that quality assurance rests first upon the adequate analytical characterization of the studied drugs. This proceeds probably from the strong tradition and value of the pharmaceutical sciences in France and has resulted in the development of very precise analytical procedures which were often considered superfluous and as too detailed by the non-French firms. Thus, many items of the rules proposed by the FDA for the test and control substances are already almost obligatory in France; a great difference is that they are not asked from the 'facility' but from an analytical expert. It is prevalent that the expert in pharmacology and toxicology must have a knowledge of the analytical report before beginning his work i.e. must know the identity, quality, purity, stability, etc. of the test substance. Furthermore, it was already stated that each batch of the substances which were handled by the expert and those which were used for comparative assays must have been analysed. These batches must be appropriately 'labelled'. Again, a difference is that then batches must not necessarily be analysed again during the experiment. Further, the French regulation always imposes that not only the drug but the formulation be thoroughly verified by the analytical expert. It is also common for the formulation itself to be studied by the pharmaco-toxicological expert. This has been later enforced by the protocol of the European Communities at least for the acute toxicity studies. This again is a characteristic of French procedure which has often been misunderstood. Its empirical basis is that some of the most tragic accidents e.g. the sulfonamid story in USA, the Stalinon one in France resulted from the toxicity, not of the drug itself but either of its particular physical presentation (sulfonamides dissolved in p.e.g. which were absorbed more effectively than in normal less soluble presentation) or of reaction of components of a mixture. I myself had opportunities to encounter cases where toxicity was either decreased (by inactivation of the drug through the pharmaceutical process) or increased (especially by different solvents). In various cases, the concentration of the drug (active principle) in the final formulation is too low to allow for the accurate determination of the toxicity; this is not only foreseeable but must be established by giving to the animals the highest dose of the formulation which can be reasonably administered. In such cases at least, it is advisable to assess the activity of the formulation and check if it corresponds to its content. Indeed, this experiment is also highly useful in any case as it is well known that some chemical modification might not modify the acute toxicity but might profoundly affect the activity. The French practice is often to perform in addition a short term toxicity study on the formulation as some adverse effects of the particular combination of chemicals cannot be disclosed by acute toxicity tests.

As in the FDA proposed rules, it is obligatory in France that the programme of the investigation be established by the manufacturer (which is the 'sponsor') and the expert but no specification is given except that this programme must comply with the legal protocol. This protocol briefly describes the classical type of assays (acute, short term and long term toxicities, fetal toxicity, fertility studies, carcinogenicity, pharmacology) and when they must be done and when they can be omitted. Various details are also given, some of which precisely illustrate the difficulty in giving technical precisions in a legal regulation; e.g. it is stated that the acute toxicity test must be carried out on equal numbers of male and female animals so that *sensu strictu*, a test might be invalidated if for example 50 males and 45 females were used.

Fortunately, many points of the protocol can be modified when the experimenter judges it proper and justifies it adequately. In this way, the protocol-orientated process does not interfere with the necessary flexibility, neither does it stifle the use of informed scientific judgement.

As required also by the FDA the final report must contain:

(a) a detailed experimental protocol giving a description of the methods, apparatus and material used, details of the species and the breed and strain of animals, where they were obtained, their number and the conditions under which they were housed and fed, stating, *inter alia*, whether they were specific pathogen free or not; omission of any of the tests listed above (acute toxicity etc.) must be explained.

(b) all the important results obtained, whether favourable or unfavourable. The original data should be described with sufficient detail to allow the results to be critically evaluated independently of their interpretation by the author.

(c) a statistical analysis of the results where such is called for by the test programme and variance within the data.

Unlike the report proposed by the Commissioner of FDA, no details such as procedure for identification of the test system, the location where all raw data are stored, etc. are asked for.

This report must also be presented as a scientific report, with an introduction, references, discussions, conclusions on the properties, safety margins, side effects, fields of application, active doses levels, incompatibilities and all information useful to the clinicians, including suggestions of possible treatment for acute toxic reactions and any side effects. In preparing such reports certain points must be emphasized.

(a) Occurrence of toxic and (or) side effects

It took some 17 years for the toxicologists to impose the view that toxicity studies—even when done for establishing safety—must disclose toxic effects of the studied drug. This is now recognized by, e.g. the protocol of the European communities which specifies 'the goal of the toxicity study by repeated administration is to show the functional or anatomo-pathological alterations and to establish their occurrence in relation to dose administered'. Later under the heading pharmacodynamics 'the experimenter must give a general characterization of the product with as a particular goal the possible occurence of side effects.'

The lack of toxicity always reminds me of some statements made about thalidomide which was presented as a non-toxic drug. More factually, it indicates often that the experiments were badly conceived or executed or that their results were falsified. If a new drug is really relatively free of some side effects commonly encountered in other drugs of the same class, such statement demands a great number of very precise studies. In these studies, adequate ranges of doses and frequencies of administration are often crucial; it might be remembered that the effects of high doses still pose some problems in teratology and in carcinogenicity.

(b) Straight mathematical handling of the results

In France, a maxim states that three kinds of lying exist which are of increasing gravity: simple lying, perjury and statistics. This is of course an oversimplification but when very complex statistical treatments are needed and used to either affirm or deny the existence of a phenomenon, be it toxic or not, the suspicion arises that there is something wrong in the experiments or interpretations. The same is true for figures when the coordinates are not easily comparable for related phenomena. Of course statistics are often necessary and have sometimes been seriously neglected. With the usual numbers of animals, it must be remembered that some statistically significant results may be quite irrelevant and that the converse is also true. Both statistics and common sense treatment of data are themselves needed.

As in other scientific studies, procedures other than mere repetition of experiences must be used to test conclusions derived only from statistical significance if this latter appears insufficient. This would be one of the goals of acute and subacute 'post screening' effects which will be discussed later.

(c) A reasonable, i.e. not excessive consistency of the results

A common saying in France is 'who would act like an angel often acts like a fool'. Results which are too consistent often indicate artificial or poor experimental conditions as do inconsistent results. I myself experienced it early in my career when I found remarkably narrow fiducial limits because the dose levels were incorrect. Too consistent results also may sometimes indicate falsification.

*(d) Presentation facilitating comparisons of detailed assembled
and summarized results as well as of detailed and
general conclusions.*

To assemble and thereafter summarize results and conclusions is a crucial
task, as indicated by the protocol of European communities. The findings
must be rendered accessible to independent persons who also can then
evaluate the data.

(e) Particular details and sometimes mechanisms of action

Details especially when repeated can correspond to aspects of the drug
or of the experimenter which cannot be obtained except when the assays
are of good quality.

Although the study of the mechanisms of a toxic action is often the
purpose of a theoretical rather than of a preclinical toxicological study,
experiments designed to assesss such mechanisms, especially when unexpected are often a sign of quality of conception and of good adap-
tation of a protocol to the obtained results. This is however not the
case for stereotyped mechanistical approaches nor for forced questionable originality.

As stated above, the rules proposed by the Commissioner of Food and
Drugs are much more detailed. From a non-regulatory point of view,
rules on organization and personnel, facilities for animal care, supply,
laboratory operation areas, specimen and data storage, design, maintenance and calibration of equipment, test and control substances, protocol
and conduct of the assays, record and reports are indeed very useful
guide-lines and also encourage improvements in experimental toxicological work.

However, the proposed FDA rules of GLP are intended to be followed by legal application and despite the great prudence and care in
their drafting, it is unavoidable that some contradictions or unforeseen
difficulties might occur. As the Commissioner has invited us to do, I will
give some examples.

It is stated that a testing facility may range down to an individual.
How can an individual be together the director of the study and the
quality insurance unit when it is stipulated that they must be distinct?
It is also stated that a testing facility can be foreign to the USA; will
the rules concerning the personnel be compatible with the local legislation? will the reporting, in the record, of a specific illness and of its
treatment be compatible with medical secrecy? The Commissioner for

Food and Drugs has stipulated fortunately the criterion for grounds of disqualification with often some modification of the rules themselves, e.g. 'and these deficiencies may have adversely affected the health of the test system or the quality and the integrity of the data'. Every experimenter knows the importance of animal care and health for good practice and the high costs of utilizing unhealthy animals; however definition of health should be given to avoid abuse and one must not forget that sometimes, accidentally ill animals can yield useful results e.g. I once noted an increased susceptibility to infection by a drug (non-corticoid) and in the clinic some related observations were done. For my part, I would think that the condition: 'affected the quality and the integrity of the results' is necessary and sufficient to cover the very goal of the proposed rules unless an undebatable definition of health is given.

But the most important and general criticism is the bias indicated by the Commissioner of Food and Drugs himself. 'These rules are a product of agency experience in reviewing long term toxicology studies in animals'. If they were to be applied as such to acute and short term studies, they might impede these latter despite their great importance.

It is needless to say that the acute studies allow the assessment of several effects and side effects, of safety margins, of species differences and of interactions between components of a mixture (including those of formulations, see p. 85) in conditions which are easy to control and inexpensive. The same is true for short term studies either pharmacological or toxicological and, e.g. short term toxicity studies of a mixture or of a formulation give results which can substitute and (or) complement pharmacokinetic studies.

Such studies can be done adequately not only in big facilities but also in smaller ones without quality assurance units and without analysts (provided analytical controls have been done by other qualified persons) and to which various attenuation of the rules could be consented without threat to the quality and integrity of the results.

Still more important is the fact that acute and short term studies are capable of revealing side effects and toxic signs for which long term studies are still often considered necessary.

This is the case e.g. for the tolerance to and dependance on narcotics which can be induced reproducibly in different animal species by single doses of such agents and within a few hours or days. Besides their heuristic value which needs no comment here and their screening value, which is self-evident for most of the participants of this meeting, they should also be given a post-screening value. Reports are rather frequent of analgesics which are declared to not induce tolerance and dependance because no such phenomena were observed in chronic direct addicting experiments in monkeys and, sometimes, in substitution studies. This unfortunately often results from inadequate dosing of the animals, adequate dosing being indeed often very difficult to establish. To verify such optimistic statements, new chronic experiments are usually done by specialized laboratories and these often disclose the production of dependence. Acute or subacute experiments can detect this property with adequate ranges of doses and time schedules much less expensively. This is only an example, but the same is true, and might become true of other acute models of chronic toxicity effects.

Such other models were commented on at the 1975 Meeting of the European Society of Toxicology where a symposium and several communications were devoted to the prediction of chronic toxicity from short term studies. *Inter alia* it was recognized that for carcinogenicity such models were still too imperfect to be applied systemically and replace the classical long term assays. However, owing to the pitfalls and the complexity of these classical long term assays, the acute models are nevertheless increasingly studied as 'post-screening procedure' and this should be encouraged, not only as independent or official research but also as non-clinical laboratories studies. In the former case, as suggested, in the FDA document, some of the rules might be used for example, allocation of grants. In the second case however, at improvement cannot proceed from the threat of legal sanction but from some recognition that such assays—if properly done and interpreted—can be a criterion of good quality and integrity of the study.

Let us hope that quality and integrity control will be adjusted in such a way that it will promote the developments of toxicology, increase the safety of the drugs and chemicals and not impede the development of therapeutics.

8

The importance of experimental design
Dennis V. Parke

The quality of the evidence on the safety and efficacy of a new drug, food additive, or other chemical, can only be as good as the overall experimental design. It is necessary, therefore, since every drug is unique, that the scientific and clinical evidence adequately demonstrates the safety and efficacy of the particular new product, and does not merely comply empirically with a series of toxicology check-lists. Moreover, data on the chemistry, pharmacology, toxicology and clinical studies should not be presented as a series of independent, and even unrelated, studies but need to be co-ordinated so as to present an integrated picture of the related safety and efficacy of the new chemical. Deficiencies in the number of animal species studied, the size of experimental groups, or the level of dosage, in metabolic or pharmacokinetic data, in evidence regarding carcinogenic or teratogenic potential, in reproduction studies, or deficiency of well-monitored, double-blind clinical trials, may delay the registration of a new drug or chemical, with the need for further studies to repair the omissions, the incurring of disproportionate costs through loss of marketing opportunity and also the loss of patient or social benefit. Advanced planning of comprehensive, integrated toxicological studies, and clinical trials, are therefore essential to the successful and rapid introduction of a new drug or chemical.

The standards expected by registration authorities are generally laid down by precedents established by the pharmaceutical or chemical industry itself, especially by the more progressive companies, and the evidence of safety and efficacy provided by many such companies is usually adequate and on occasions extremely thorough. However, there are other companies that produce inadequate data or naive interpretations, and still others which do not yet see the need to provide such evidence, often because the drug is a naturally-occurring substance or has had prolonged human use elsewhere.

NATURE OF DRUG SUBSTANCE

The identity, degree of purity, nature of impurities, radiochemical purity where isotope-labelled materials are being used, and particle size of the drug substance are of paramount importance in any experiment, and especially in the evaluation of toxicity. Yet very few published papers, let alone new drug submissions, give adequate data concerning these aspects. With modern analytical techniques, such as high pressure liquid chromatography and gas liquid chromatography-mass spectrometry, trace impurities may be separated and identified unequivocally, and then quantified with an extremely high degree of sensitivity.

Several instances are known where a drug synthesized by one route has been shown to exhibit toxic characteristics at high dosage not seen in the same compound prepared by an entirely different route. Presumably these differences in toxicity observed in two different preparations of the same chemical must be due to toxic trace impurities, even though both preparations may have been shown to be >95% pure. Similarly, the method of preparation of natural products may affect their purity and hence their therapeutic activity, for example, insulin which may contain varying amounts of glucagon.

Frequently the preliminary toxicity evaluation may be carried out using one chemical form of the drug, say the free acid, only to find that it is desirable to administer the drug to patients as a salt. This presents the toxicologist with the dilemma of whether or not it is necessary to repeat all of his studies on the new form of the material. Often it is assumed that there will be no difference in toxicity, but there have been cases where the free acid and the sodium salt have been shown to be non-toxic yet the salt with an organic base, such as choline, was significantly hepatotoxic. Presumably, this again was due to trace impurities present in the organic base.

The particle size of a drug may profoundly affect the rate and extent of absorption of a drug, especially when administered orally, and may thus affect the time and magnitude of the peak plasma concentration, the area under the curve, and the degree and duration of the pharmacological activity. For these reasons it is imperative to undertake toxicity evaluation on material of the same particle size as intended for administration to patients.

A similar problem is the evaluation of delayed release preparations, especially when a normal release preparation is already on the market. A comparison of the plasma pharmacokinetics of the two preparations will provide sufficient evidence of comparable efficacy and safety only when it is known that the therapeutic effect of this particular drug is directly related to plasma concentration. Where this is not so, additional animal toxicology and human clinical evaluation may be necessary.

MODE OF ACTION

With the introduction of any new pharmacological principle it is desirable to have an adequate understanding of the mechanism(s) involved. In this way, potential areas of toxicity may be predicted, and safety evaluation may be appropriately designed to monitor these risks. For example, anti-hypercholesteraemic agents may reduce the blood cholesterol by (a) changing the tissue distribution of the sterol, (b) inhibiting absorption of exogenous cholesterol, (c) inhibiting endogenous cholesterol biosynthesis, (d) accelerating cholesterol catabolism, or (e) increasing the excretion of bile acids by lowering their enterohepatic recirculation. Each different mode of action may involve different associated phenomena, which may lead to entirely different toxic manifestations and hence would require different monitoring. Thus, if the anti-hypercholesteraemic agent acts by changing the tissue distribution of the sterol it may be most appropriate to monitor liver function and cholesterol content, whereas if the drug acts by increasing the faecal excretion of bile acids it would be more appropriate to monitor gastrointestinal malabsorption of acidic substances such as ascorbate, folate, and acidic drugs.

Similarly, drugs which block β-adrenoreceptors and combat hypertension may also result in other seemingly unrelated biochemical effects, such as impairment of mucus synthesis (Parke and Symons, 1977); the new H_2-receptor antagonists used in the treatment of hypersecretion of gastric acid, might also affect H_2-receptors in other organs and tissues, such as lymphocytes (Gomard et al., 1976).

ANIMAL TOXICOLOGY

Many deficiencies in animal toxicology data are due to inadequate planning of the experimental studies. Toxicology studies have frequently been invalidated by the use of unhealthy stock, because of inadequate planning to ensure the breeding or purchase of disease-free specimens or to effect the eradication of parasites; by the use of insufficiently high dosage, probably because the toxicologist was misled by incorrect information as to the likely dosage to humans, or because of inadequate acute toxicity ranging studies; and by the lack of post-mortem examination of animals dying during the night. The use of healthy stock, and good animal husbandry, is too important a subject to do justice to in this brief survey. Yet so often unsatisfactory data is explained away as due to the use of unhealthy animals, as though this were some misfortune over which the experimenter had no control. Attention to these fundamental principles of good biological experimentation is of the greatest importance.

Acute toxicity studies, especially the determination of LD_{50} values, have recently received much public criticism, largely because of the usual inelegance of the experimental design and high cost in animal lives (Sperling and McLaughlin, 1975). Post-mortem examination of animals which die, and establishment of the cause of death, with possible elucidation of the mechanism of toxicity, would provide invaluable additional information for establishing suitable dosage levels and appropriate monitoring for chronic toxicity studies. As animals are usually under experimental observation for only eight hours per day it should be expected that when the death of an animal occurs it is likely to be during the night, often with loss of post-mortem opportunity because of autolysis or cannibalism. Arrangements should be made to deal with this eventuality, and one progressive company, at least, has solved the problem by having its night security staff inspect all experimental animals hourly and remove any dead to the refrigerator to await post-mortem examination next morning.

One last matter concerning chronic toxicity studies is the frequency of dosing. Presumably for reasons of staff convenience, some toxicology is still being conducted with administration of the drug on 5 days only per week, with or without double dosage on Mondays and Fridays. Recently a company made a plea for the scientific appropriateness of this 5-day regimen of convenience, because the material under test was an industrial chemical to which people were exposed only when at work, that is, on Mondays to Fridays. The argument was accepted, but it was suggested by the regulatory authority that studies involving both 5- *and* 7-day dosage regimens were desirable.

CLINICAL BIOCHEMISTRY

The data produced by chemical analysis of the urine and blood of experimental animals and human patients, comprises a plethora of numbers concerning concentration of the drug, electrolytes, enzymes and other biological constituents. Many of the numbers are print-outs from computers linked to autoanalysers, and give no indication of statistical, methodological or biological significance. The incorporation of this print-out data without correction for the limitations of the methodology can give erroneous impressions of the sensitivity of the method and the significance of the results, which may be as misleading to the industrial toxicologist as to the assessor for the regulating authorities. For example, the computer print-out of plasma drug concentration data shown in Table 8.1 give grossly exaggerated impressions of the accuracy and sensitivity of the method, which on closer examination showed that the instrument readings were below the optimum range of the instrument, and consequently the data were invalid. Furthermore, close examination of such data, as by number frequency, has been known to reveal inconsistencies which have been used by some companies to identify unreliable sources of data. Table 8.2 shows incorrectly presented data concerning serum cholesterol concentration, which was validated by examination of the original print-out data and then corrected for the accuracy of the chemical method.

TABLE 8.1   Accuracy and sensitivity of plasma drug concentration data

| Sample No. | Plasma drug concentration (μg/ml) | | Instrument reading |
	Computer print-out	Corrected	
1	5.526	5.5	0.09
2	4.298	4.3	0.07
3	4.912	4.9	0.08
4	4.298	4.3	0.07
5	6.754	6.8	0.11
6	5.526	5.5	0.09
Accuracy:	$\pm 0.1\%$	$\pm 1\%$	$\pm 10\%$
Sensitivity:	0.001 μg/ml	0.1 μg/ml	1 μg/ml (0.02)

TABLE 8.2   Validation of clinical biochemistry data

	Serum cholesterol concentration (mg/100 ml)		
	Presented data	*Print-out data*	*Corrected data*
	270.14 ± 23.14	269.14 ± 23.140	270 ± 23
	245.47 ± 17.62	246.47 ± 17.627	245 ± 18
	260.35 ± 28.37	258.35 ± 28.370	260 ± 28
	220.63 ± 12.45	222.63 ± 12.458	220 ± 12
	280.22 ± 18.31	280.22 ± 18.319	280 ± 18
	225.14 ± 26.36	223.14 ± 26.362	225 ± 26
Frequency of last 3 digits	0(4) 1(4) 2(4)	0(3) 1(4) 2(5)	
	3(6) 4(5) 5(4)	3(6) 4(5) 5(2)	
	6(4) 7(3) 8(2)	6(4) 7(3) 8(2)	
	9(0)	9(2)	

In selecting biochemical parameters for monitoring organ function, or rather, organ integrity, care should be taken to choose appropriate enzymes, etc. The standard parameters used in diagnosing natural disease may not always be as appropriate for monitoring drug toxicity. Many new drugs are highly bound to plasma proteins, are largely excreted in the bile, and lead to induction of hepatic enzymes. As a consequence they may result in slightly raised serum bilirubin concentrations, and raised levels of serum alkaline phosphatase, γ-glutamyl transpeptidase and the transaminases. Having selected these parameters for monitoring liver function, when they are subsequently found to be raised, it is unethical to dismiss them as of no pathological significance simply because it is believed that other known non-toxic effects of the drug are responsible for the abnormal values (see Table 8.3). This is a problem with several new drugs and we are at present carrying out studies to find more satisfactory ways of monitoring drug hepatotoxicity.

PHARMACOKINETICS

The real aims of pharmacokinetic and metabolism studies often seem to be overlooked, with the result that studies presented are often incomplete and inconclusive. The essential purpose of such studies is to establish the degree of absorption of the drug, its routes and rate of elimination, the nature of chemical changes occurring to the drug, and the temporal distribution of the drug and its metabolites in the body tissues,

TABLE 8.3 Drug-induced reversible changes in liver function tests

Plasma enzyme activity	Possible reasons for increases
Alkaline phosphatase ↑	Competition for biliary excretion by drug
Oxalacetate aminotransferase ↑ (SGOT)	Enzyme induction (non-specific) by drug
Pyruvate aminotransferase ↑ (SGPT)	Enzyme induction (non-specific) by drug
γ-Glutamyl transpeptidase ↑ (γ-GT)	Enzyme induction (glutathione conjugation of drug)
Lactate dehydrogenase ↑ (LDH)	Drug modifies membrane permeability
Serum bilirubin ↑	Displacement from plasma protein binding by drug

especially those involving the target(s) of the drug action. By comparison of pharmacokinetics in different animal species, these metabolic studies also serve to validate the choice of species selected as a model for man in the toxicology studies.

Induction of the hepatic microsomal enzymes, which occurs with many drugs especially those which are slowly metabolized, is regarded by some as implicit of toxicity. This is not necessarily so, although long-term administration of drugs which are enzyme inducing agents may result in secondary effects on certain aspects of intermediary metabolism, resulting in folate depletion and possible teratogenesis, vitamin D deficiency, osteomalacia and possibly osteoporosis. For these reasons it is desirable to obtain some knowledge of the enzyme-inducing potential of the new chemical.

Recent studies in our laboratories have shown that with both animals and man, folate deficiency is dependent on the adequacy of the diet and on the duration of administration of the enzyme-inducing drug; it does not appear to be related to the chemical nature or to the pharmacological properties of the drugs *per se* (Labadarios, 1975). Table 8.4 shows the effects of chronic treatment with various drugs on the serum folate concentrations, urinary formiminoglutamic acid (FIGLU) excretion, and microsomal enzyme induction (cytochrome P-450) of rats maintained on folate-deficient and folate-supplemented diets. Table 8.5 shows the reproductive effects of the treatment of rats with anticonvulsant, enzyme-inducing drugs (phenobarbitone plus diphenylhydantoin)

TABLE 8.4   The effects of drugs and folate supplementation on folate status in rat

Treatment	Folate-supplemented diet			Folate-deficient diet		
	Serum folate (ng/ml)	Liver Cyt.P-450 (nmol/g liver)	Urine excretion of FIGLU (mg/h)	Serum folate (ng/ml)	Liver Cyt.P-450 (nmol/g liver)	Urine excretion of FIGLU (mg/h)
Controls	20.6±0.2	18.2±0.6	0.1±0.01	6.0±0.4	19.2±1.4	0.5±0.02
Phenobarbitone	20.8±0.2	49.0±3.0	0.1±0.01	4.3±0.2	20.0±1.7	8.6±0.1
Diphenylhydantoin	20.5±0.2	22.0±0.9	0.1±0.01	4.3±0.4	19.6±0.9	1.2±0.1
Imipramine	20.4±0.2	29.0±2.0	0.1±0.01	4.3±0.5	15.0±0.9	1.3±0.1
Succinyl sulphathiazole	18.0±0.3	24.1±0.6	0.15±0.01	2.5±0.1	13.6±1.5	1.3±0.1

Note decreased serum folate, increased urinary FIGLU excretion and absence of induction of cytochrome P-450 in animals maintained on folate-deficient diet.

TABLE 8.5 The folate status and fetal weights of rats maintained on folate-deficient diets with and without anticonvulsant drug administration

Diet	Mean fetal weight (g)	Mean litter weight (g)	Maternal serum folate (ng/ml)	Maternal erythrocyte folate (ng/ml)
Folate deficient:				
Control	2.2±0.3	22.9±4.1	5.8±6.7	50±1
Plus anticonvulsants	1.9±0.2	13.6±2.5	3.1±0.2	34±3
Folate semi-supplemented:				
Control	2.8±0.3	35.6±5.1	7.4±0.4	70±3
Plus anticonvulsants	2.4±0.3	30.6±0.3	6.0±0.5	58±4
Folate full-supplemented:				
Control	3.5±0.3	28.0±3.6	11.4±1.3	89±7
Plus anticonvulsants	3.4±0.1	38.7±3.9	8.3±0.1	72±7

for one month before mating and throughout pregnancy, when maintained on different folate-containing diets. The effects of the folate deficient diets and anticonvulsant drugs on fetal development are shown in Figure 8.1. Folate deficiency leads to skeletal abnormalities which are exacerbated by the administration of the enzyme-inducing anticonvulsant drugs. It is therefore essential, when carrying out toxicology and reproductive studies with enzyme-inducing drugs, to ensure that the diet provides for an adequate intake of folic acid.

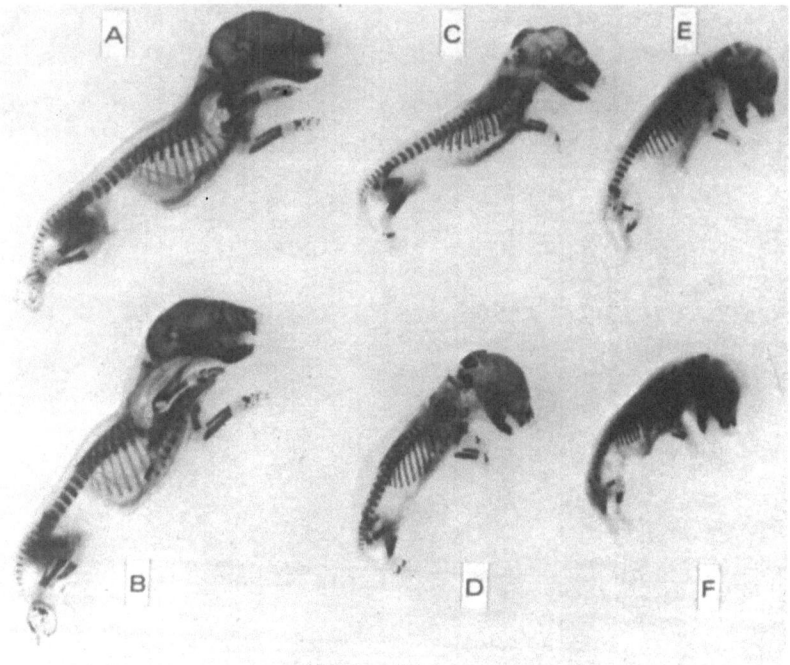

Figure 8.1 Skeletal malformations in neonatal rats born to dams maintained on various folate-containing diets and given anti-convulsant drugs. Dams were maintained on the various folate-containing diets for one month before mating and throughout pregnancy, and were given phenobarbitone plus diphenylhydantoin daily at the equivalent human therapeutic dosage. A, normal diet, no drugs; B, normal diet + anticonvulsant drugs; C, folate semi-supplemented diet, no drugs; D, folate semi-supplemented diet + anticonvulsant drugs; E, folate-deficient diet, no drugs; F, folate-deficient diet + anti-convulsant drugs. Note reduction of fetal size, shortening of bones, broken ribs, skeletal defects and impaired calcification in D, E and F.

Because of enzyme induction, and also because of bilary excretion and enterohepatic recirculation of the drug, the pattern of metabolism and pharmacokinetics may change radically after repeated administration. This can lead to changes in the patterns of toxicity and pharmacological activity, and may also lead to progressive accumulation of the drug in the tissues. It is therefore essential that the pattern of metabolism and pharmacokinetics of a drug be determined in animals and man, not only after single dosage, but also after repeated dosage over a period of about one month.

CARCINOGENICITY AND MUTAGENICITY

The need for carcinogenicity studies varies greatly with different regulatory authorities. Since it is not possible, with any degree of certainty, to predict from chemical structure, or pharmacological activity, whether or not a chemical is a potential carcinogen, one could argue that all new drugs should be examined for this possibility. Furthermore, in this situation of uncertainty, it is surely unethical to expose even one patient or one volunteer to this risk. But this rather extreme view would mean undertaking carcinogenicity studies before clinical trial, before clinical pharmacology, and even before administration to healthy volunteers. And if such studies were to be undertaken, would a study in rodents be adequate (they were not for the oral contraceptive steroids), or would studies in dogs or even primates be necessary?

Although short-term, *in vitro* tests for carcinogenicity are still undergoing evaluation, they may offer some solution to this problem. Not all mutagens are carcinogens, but several workers have shown good correlation between a positive Ames test and the carcinogenic potential of chemicals (Purchase *et al.*, 1976). Several companies are now providing evidence of the reactions of their new drugs and chemicals in a battery of these short-term tests, as some evidence of safety. In combination with other evidence, such as efficient and rapid metabolic detoxication and excretion, and absence of covalent binding and tissue accumulation, these tests go some way, at least, to providing an assurance of lack of carcinogenic potential.

REPRODUCTIVE STUDIES

It has been said that all chemicals can be teratogenic or fetotoxic provided that they are administered at sufficiently high dosage. Indeed, it has been observed that compounds known to be free from teratogenic effects when administered normally, even at high dosage, can result in fetal abnormalities when administered rapidly by intravenous injection. The result of such a bolus injection is a massive transplacental transfer of the drug, often several orders of magnitude greater than would occur following slow intravenous infusion of the same dose. This phenomenon has given rise to several instances of spurious indications of fetotoxicity and teratogenicity when intravenous administration has been made by an inexperienced operator.

There is a tendency to defer reproductive and teratogenicity studies to the final stages of the toxicity evaluation of a drug, with the suggestion that women of reproductive potential be excluded from clinical trials. Reproductive studies, however, yield information concerning aspects of toxicology other than 'fetotoxicity' or impairment of reproductive potential. Several workers have shown a correlation between teratogenicity and carcinogenicity, and Figure 8.2 shows this relationship in different organs of rats exposed to the carcinogenic triazenes at different periods of fetal development (Druckrey, 1973). Reproductive and teratology studies should therefore be an integral part of the comprehensive evaluation of potential toxicity, and should be completed before any extensive administration of the drug to humans.

CLINICAL TRIALS

Although the purpose of a clinical trial is primarily to establish the efficacy of a new drug, this is the first real opportunity to evaluate its safety in the relevant species, namely, man. In many cases the toxic side-effects of a drug become manifest only after extensive use of the drug, and sometimes only after prolonged use as well. For unlike the laboratory animals with their high degree of genetic homogeneity, their relatively young age, and their freedom from disease, the patients who receive these drugs are genetically heterogeneous, often relatively old, and presumably sick. It is not unreasonable, therefore, to expect the appearance of adverse side-effects when administering a drug to this heterogeneous, sick population. The ideal situation prevailing in the laboratory cannot, because of the very circumstances of disease, be matched when one takes the drug to the patient.

Hence, it is even more imperative to carefully plan the experimental studies in man. Adequate monitoring for the toxic effects seen in the experimental animals receiving high dosage of the drug is essential. Moreover, the optimum safe dosage should be determined not once for

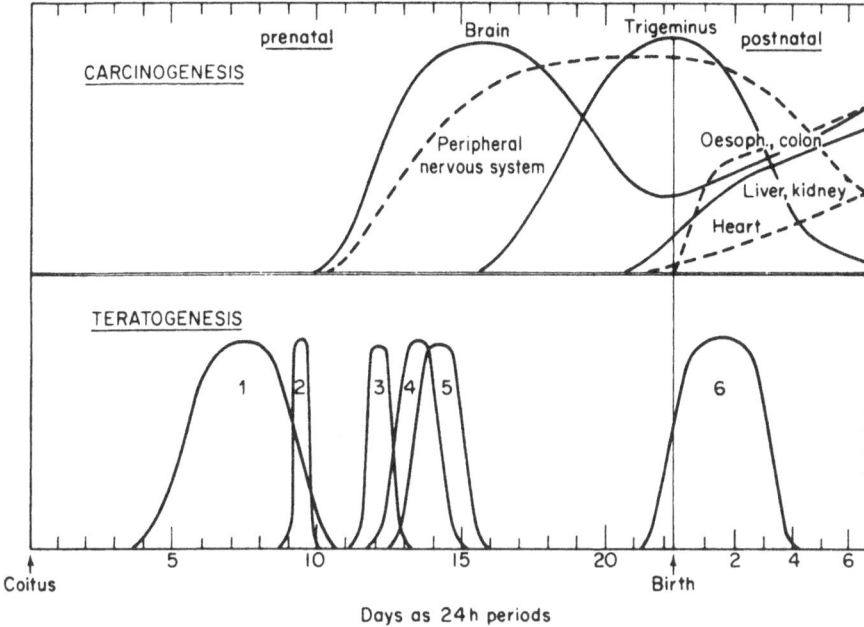

Figure 8.2 Sensitive periods of various organs of rats in teratogenesis transplacental and neonatal carcinogenesis. Malformations induced during the periods: (1) brain—lethal doses; (2) optic nerves and eyes—at high doses; (3) fore-limbs, skeleton; (4) hind limbs, skeleton, cleft palate; (5) microencephalia, (6) ears. (From H. Druckrey, *Xenobiotica*, 3: 271 (1973))

all time, but repeatedly throughout the treatment, checking by determination of plasma concentrations of the drug and by re-evaluation of the pharmacokinetics after a period of prolonged administration, to ascertain that the dosage is appropriate to obtain maximum efficacy compatible with adequate safety. Moreover, because of pharmacogenetic differences, dosage for such drugs may even need to be individually determined. As an extension of this, one can see the need for clinical trials in the type of population that will receive the drug. Because of differences in pharmacogenetics, diet, social customs and medical practice, dose levels evaluated in Japan may not be suitable for Europeans and *vice versa*. Ultimately, one must remember that modern efficacious drugs are frequently selectively toxic agents, excessive administration of which may be harmful. Hence, the optimum safe dosage may be more critical to the patient than is generally assumed.

PRESENTATION OF EVIDENCE

Finally, to return to my original theme, the evaluation of safety and efficacy of a drug, or other chemical, requires the integration of all the different aspects of the study, and a co-ordination of the interpretations. Too many presentations of safety/efficacy data are obviously the work of several individuals with little or no co-ordination of the different results. Indeed I have read presentations where frank conflict was apparent in the interpretation of results and opinions of significance expressed in the different sections of a new drug submission.

In order to achieve this co-ordination it is highly desirable that the entire presentation should be finalized by one individual, or at the most two, say a clinician and a toxicologist. These two individuals should accept final professional responsibility for the whole submission and be capable of elucidating any obscure points to the regulating authorities. This would be the ultimate means of quality control, and would mean a return to true professionalism which would go a long way to dispelling the increasing mistrust which is apparent between industry, government and the medical profession. At a recent international meeting of toxicologists from industry, regulating authorities and academia, the evaluation of drug toxicity exclusively 'in house' was regarded by most with some suspicion and it was considered desirable that at least some part of the evaluation should be conducted in independent laboratories. This too, might well increase the confidence in the overall study, and is indeed already the practice of many of the major industrial companies.

References

1. Druckrey, H. (1973). Specific carcinogenic and teratogenic effects of 'indirect' alkylating methyl and ethyl compounds, and their dependency on stages of ontogenic developments. *Xenobiotica*, **3**: 271
2. Gomard, E., Duprez, V., Henin, Y. and Levy, J. P. (1976). H-2 region product as a determinant in immune cytolysis of synergeic tumour cells by anti-MSV T lymphocytes, *Nature, Lond.*, **260**: 707
3. Labadarios, D. (1975). Studies on the effects of drugs on nutritional status. (PhD Thesis, University of Surrey)
4. Parke, D. V. and Symons, A. (1977). The biochemical pharmacology of mucus, in *Mucus in Health and Disease*, (M. Elstein and D. V. Parke, editors) Plenum: London
5. Purchase, I. F. H., Longstaff, E., Ashby, J., Styles, J. A. and Anderson, D. (1976). Evaluation of six short-term tests for detecting organic chemical carcinogens and recommendations for their use, *Nature, Lond.*, **264**: 624
6. Sperling, F. and McLaughlin, J. K. (1975). Biological parameters and the acute LD_{50} test, *J. Assoc. off Anal. Chem.*, **59**, 734

SECTION FOUR

The Research Contractor
Chairman:
J. McL Philp

9

Quality assurance in contract research organizations
Peter R. B. Noel

INTRODUCTION

The *quality* of a 'thing', such as we are concerned with today, that is, the performance of scientific work, is defined in the Shorter Oxford English Dictionary as 'an attribute, property or special feature'. It was only about 100 years ago that the definition changed to include an element of 'peculiar excellence or superiority'. Today, therefore, the most appropriate definition would seem to be:

The *degree or grade of excellence associated with, that aspect of things under which they are being considered, in thinking or speaking of their nature, condition or properties.*

In scientific terminology, characteristics can be described either qualitatively or quantitatively, the difference being that the latter is measurable, whilst the former is not. From the definition, it follows that *quality assurance* and *quality control* are meaningless phrases unless there is some *standard against which a comparison can be made*. In the United Kingdom it has, to date, been rare for written standards of quality to exist: instead, the principle has evolved whereby when a qualitative aspect is in doubt, a comparison is made against the performance of one's peers. Until called upon to account for one's actions, the operating system which involves two parties, is basically one of MUTUAL TRUST.

Such a situation and method of operation has tended to exist throughout the world as regards the performance of scientific work. It is for this reason that in the realm of assessment of the safety of chemicals recent investigations which have resulted in the description of deficiencies in technique, uncovered by the United States FDA, have made such depressing reading because they represent a breach of existing trust. Against such a background, therefore, one may fairly conclude that the FDA had no practical alternative to the production of its 'Proposed Regulations for Good Laboratory Practice'.

Governmental agencies have a responsibility to the public to review the results of scientific work and make a decision as to whether, in their opinion, a particular chemical should be introduced into the environment of the country for which they are responsible. It is obviously essential, therefore, that they are in a position of trusting the accuracy and truthfulness of the documentation placed before them. When this trust no longer exists, what alternatives do they have other than the introduction of some checking system? Having decided to introduce independent checking, they then have to decide whether to adopt an isolationist policy (only to accept data produced within their own borders, under their direct control) or to continue as at present and ensure, directly or indirectly, that the same controls exist beyond their boundaries and that they are allowed to check that this is so. It seems that the FDA have decided to take the latter course and achieve worldwide control through the sponsor-testing facility relationship indicated in the Proposed Regulations.

A testing facility may be either part of a sponsoring company's own organization or independent. Contract research organizations, especially those in the UK, fall into the category of *consultancies*, which implies that they have independence and a degree of expertise. The consultative activity may be associated with the performance of some work, all of which is financed by fees. The consultancy element may be evident in the planning stages of a study (production of a protocol) or during the course of a study, but it is always present at the termination of a study, when the data obtained has to be interpreted. It has always been my belief that such an independent consultative attitude exists, not only in the UK, but also throughout the rest of the world and, if so, implies that, for example, contract research organizations must make themselves particularly attractive to sponsors through the maintenance of high standards of quality.

FDA PROPOSALS

The principal concern of this meeting is obviously the document that appeared in the Federal Register, Friday, 19th November 1976: Part II. Nonclinical Laboratory Studies: Proposed Regulations for Good Laboratory Practice. My personal opinion is that the whole article is eminently reasonable, to the extent that I have yet to meet anyone who does not agree with it in principle. There is, however, a difference between acceptance in principle and acceptance in practice.

The documented proposals are reasonable, logically presented and are also very readable, so much so that it was not until my second reading that I realized almost 15 pages had been used as an explanatory preamble, followed by only 8 pages of 'Proposed Regulations' (associated with an additional couple of pages (Part 8) dealing with slight elaborations for specific chemicals). It must be remembered, therefore, that as things stand at the present moment, *only the 8-page section will become law in the United States* and thus affect sponsors and testing facilities in the United States and elsewhere. One has a tendency to read the Regulations with the words of the preamble in mind, perhaps in a similar way that one might read proposed UK legislation with the idea of justice in mind. Unfortunately, however, the two parts are distinctly separate, as is indicated in the law courts: It is only the law, in this case the 8-page regulations, which can be considered.

PRINCIPAL DANGER OF GLP

Whereas one may be sympathetic to, and even agree with, the action taken by the US Governmental Authorities, one must remember that it will be practical application of the legalized regulations that will become important in the future. With this in mind, therefore, a brief analysis of the proposed GLP Regulations is worthwhile. They consist of both definite statements and imprecise statements. The latter may be subdivided according to whether they include alternatives, as for example '. . . animals of the same species, but used in different studies, should not ordinarily be housed in the same room . . .' or require comparison to be made before determining 'adequacy'.

In addition to the GLP, most of us will have seen the 'Compliance Program Guidance Manual' dated 14th November 1976, which is to be used by Task Force inspectors during the pilot investigatory phase. Regretfully, this document in no way clarifies the imprecise statements.

Both documents use similar words, one as a statement and the other as a question, thus:

(i) GLP: The . . . shall be 'adequate'.

(ii) CPGM: Is the . . . 'adequate'?

If such terms are found in the final regulations some definition must be sought. Examples are:

sufficient	e.g. (29c) . . . shall be a *sufficient number* of personnel
	(43a) . . . shall have a *sufficient number* of animal rooms
adequate	e.g. (90f) . . . *adequate differentiation* by space and . . .
	(61a) . . . shall be of *appropriate design* and *adequate capacity* . . .
periodically	e.g. (90i) . . . shall be analysed *periodically* . . .
	(113a 1) . . . a *periodic* check of uniformity . . .
complete	e.g. (33b 4) . . . perform a *complete* evaluation . . .
each phase	e.g. (33b 3) . . . inspect *each phase* of a . . .
timely	e.g. (29c) . . . personnel for the *timely and proper* conduct of the study . . .
also	e.g. (29a) . . . shall have *education, training and experience* . . . to *enable that person to perform* the assigned functions
(also, in reverse)	(202a) . . . (disqualification) . . . personnel . . . inadequate in number or insufficiently trained . . .

This brings me almost full circle, to the point where words have to be defined in order to ensure uniformity in either qualitative or quantitative comparisons or measurements. If definitions or explanations are not included in the Proposed Regulations, then we must recognize that assessments of the way in which studies are performed will still be based on a large element of TRUST. There is a difference, however, in that this time, it is not only the FDA who have to do the trusting; it involves the sponsors and other research contractors, all of whom will have to trust the Task Force Inspectorate.

Prior to the concept of producing GLP regulations, reputable organizations always knew that if the occasion arose, they would be judged by their peers; in other words, scientific work would be judged by scientists, etc. The Proposed Regulations, as they stand, imply that scientists will now be 'judged' by non-scientists. Let me stress that this is not stated in the Regulations but is an inevitable conclusion.

Provisions are made for the eventual sorting out of any apparent differences (3e. 204 (b)), but apparently this is only possible after 'the Associate Commissioner for Compliance' issues 'a written notice proposing that the facility be disqualified', that is, before 'a hearing on the disqualification should be conducted'.

We have, in effect, then, what appears to be *a principle of having to prove one's innocence* (before the public?) rather than the more generally accepted reverse principle. This, I suggest, is the main danger, of concern to all sponsors, contract research organizations and other forms of research contractor. It means that publicity can result through misunderstandings by a non-scientific inspector, about a scientific subject, and the result would be damaging whatever the outcome: 'there is no smoke without fire'.

POSSIBLE SOLUTIONS

(i) One answer is: to define all imprecise words. This, however, would produce its own dilemma, in the form of

(a) great rigidity of the Regulations;

(b) increased cost, because of the necessity of erring on the cautious side;

(c) a legalistic approach to possible loopholes, and

(d) difficulties in that definitions would need to have worldwide meaning.

(ii) The alternative would be: to preserve the spirit of the preamble and introduce into the Regulations one of the two following requirements:

(i) All inspection teams must be headed by a scientist or, perhaps two scientists who should be in agreement, before any suggestions regarding disqualification could be made, or

(ii) there should be an intermediate stage, whereby discussions at a scientific level would have to take place with the Governmental Authority before the Associate Commissioner for Compliance was allowed to proceed along the lines suggested in the present proposals.

To preserve the spirit of the preamble, I am certain would need an expansion of the various sections of the presently Proposed Regulations, going at least some way towards definition, perhaps, for example, as to how 'adequacy' should be determined.

CONTRACT RESEARCH ORGANIZATIONS

Huntingdon Research Centre (HRC) of which I am an employee, has been established for over 25 years and for the last 15 years or more has been conducting non-clinical laboratory studies, the reports of which have been presented to governmental authorities, including the FDA. Over the latter period, we have grown from about 30 to almost 700 in staff complement, as a result of what I believe to be a reputation for integrity (*trust*worthiness) and our attempts to maintain high quality.

Of course, we are paid for the work that we do. Over the years, we have had one bone of contention for which we have never found a satisfactory solution. Not uncommonly, we are asked: 'Will you please prepare a protocol and estimate of cost for a 3-(6- etc.) month study in rats (dogs, primates, etc.) on a drug (pesticide, food additive, etc.)?' We have learned that however precise and detailed our protocols, it is the estimate of cost alone which is occasionally the basis for selecting a testing facility. Lower costs have not infrequently been reached by abbreviating protocols and sometimes, sponsors could not, or would not, appreciate the differences in the contents of the study proposed. The introduction of financial considerations leads to competition.

There is nothing wrong with competition, provided it is not allowed to affect the quality of the scientific work performed. It is incumbent upon organizations, especially those involved in contract research, to ensure that they are working efficiently as regards space, apparatus and personnel. Further reductions, however, to achieve cost cutting must be undesirable. If the GLP regulations are, therefore, to state what are the minimum standards for adequate space, numbers of staff, etc., we may at last be near to solving our particular bone of contention.

The GLP regulations are aimed principally at recording techniques. These are a form of quality control. There is little present which is applicable to true quality assurance. That the work performed is of high scientific quality is not the subject of the GLP regulations; it is presumably 'assumed' from the single requirement, specifying that personnel involved in the study should be appropriately qualified and/or experienced.

There is, I believe, now a case for reviewing other written guidelines produced by the FDA, which relate to the scientific procedures for performing adequate non-clinical laboratory studies. I would like to see a reconsideration (with justification) of some of the generalizations relating to the performance of animal studies, especially those relating to the distinction between the two objectives:

(i) The assessment of the degree of *safety* associated with the test compound, the information pertaining to which is usually associated with low, and perhaps intermediate, dose level groups of animals, and

(ii) demonstration of *possible toxicity* that might result from administration of the test compound, a feature of high dose level study where the dosage given is not necessarily related to human intake.

In the GLP preamble, there seem to be suggestions that the two objectives are basically the same. The demonstration of toxicity is relatively easy. Assessment of safety, however, which I suggest is of greater importance, is more difficult since it depends on demonstrating the absence of toxicity at levels of the chemical equivalent to maximum human intake. There is no need to dwell on the difficulties of proving a negative. When no abnormalities appear to be present in animals receiving a low dose level, it is human nature to pay less attention to that group and give more to groups producing interesting changes. It is common practice in rodent studies to perform investigations in the first place on high dose level and control animals, assuming that if no changes are seen at the high level, then none can result at lower dosage levels. Such an attitude does not help to prove that at a particular point in time those low dose level animals were normal.

The argument used seems to be logical but it must also be remembered that clinical observations, bodyweights, etc. are reported for all groups and in non-rodent studies, it is accepted practice to investigate all groups on each occasion. Obviously, cost is partly responsible for this practice but I would like to know whether *FDA scientific guidelines* would suggest:

(i) examination of samples of all (or only some, as at present) rodent groups whenever investigations are made and

(ii) the recommended method of arriving at the most satisfactory sample size.

HRC AND THE GLP

(a) Quality assurances unit

Experience at HRC has shown that, when an organization is small, it is possible for those in charge of departments to know the strengths and weaknesses of all individual graduates, who control the day-to-day running of studies. With growth, however, this personal knowledge becomes less certain and alternative controls have to be introduced.

The GLP regulations have extended this line of reasoning and suggested the introduction of internal QA units, operating independently of the scientific community but reporting to the same management. However unusual such a suggestion may seem to scientists, it does appear to be a logical and perhaps appropriate answer to the present problems. In its preamble, the FDA have questioned the applicability of such a unit to small organizations. Since a QA unit may consist of only one person, and since operation based entirely on trust is now under suspicion, I suggest that the principle should be applied to all testing facilities.

Unique to contract research organizations, is a problem being generated by the indefinite wording of parts of the GLP. After setting its own standards of quality, its QA unit may have to ensure compliance with, not only the internal standards, but also those of individual sponsors. If the latter are unusual, they may disturb smooth running and tend to defeat the GLP objectives.

It has been suggested that the operation of internal QA units would be more successful if performed by non-scientists. Since the unit's principal function is comparison between written documents: those that state 'what should have been done' as against 'what was done', I would personally agree with this suggestion and it is this method of operation that is to be first investigated by HRC. We do, however, differ in that we are:

(i) using scientists to set up the QA unit; write its own particular Procedural Manual and determine how most appropriately comparisons can be made, as well as

(ii) ensuring that the Unit will be subsequently in the charge of, and reporting to, scientists. The purpose here is to correct minor discrepancies without involving management and also to distinguish more important deficiencies, which may have an adverse effect in the interpretation of results obtained, which must be dealt with by management.

(b) Scientific departments

The desirability of having all-encompassing manuals describing the procedures associated with each department, has long been recognized. Information has, of course, always been available to ensure smooth running of the departments but, to date, this has not been collected into a single entity and, therefore, it was not as accessible as it should have been.

The GLP requirement to produce procedural manuals will, therefore, be advantageous to us, as it will also be a stimulus to review, and if necessary revise, present recording procedures.

In the GLP, emphasis is placed on *training*. This word always seems to conjure up the acquisition of knowledge, by graduates and technicians, shortly after their arrival. A problem encountered in all testing facilities is that staff do leave and it is necessary to ensure continuity, as well as the future for that organization, with the result that either there is a continual recruitment programme or a continuous training process, allowing existing staff to replace those leaving.

In HRC we have emphasized the training aspect. Since education and training to gain experience are achieved primarily on a personal basis, some time ago we introduced regular, internal meetings at two levels:

(i) Those controlled by Experimental Supervisors—to ensure that all graduates and technicians concerned with a study are aware of its objectives and the progress being made.

(ii) Those controlled by the Head of the Department for his Experimental Supervisors—not only for the purpose of updating himself, but also to train his graduates in interpretation, evaluation of data and clarity of reporting.

(c) Archives

Having produced accurate and complete records, it is logical to require their preservation, at least for a period of time. As regards the GLP Regulations, however, the situation relating to contract research organizations is slightly different from that of sponsors and their own testing facilities, at least in Europe. We at HRC have taken *legal opinion* and learnt that, although there is no specific UK law requiring retention of records for any specified period of time, there are aspects of common law which are applicable to *consultancies*.

As a result of this legal advice the following conclusions have been reached:

(i) It is the responsibility of HRC *to retain original data or legally acceptable copies of original data*, to the extent that, at a later date, they could produce evidence to confirm statements made in reports issued by HRC.

(ii) This means that, if original data is handed to a client ('sponsor') and copies are retained within HRC, then legally acceptable records should be retained to record this transfer of material and include a proviso that, before it is destroyed or transferred to any other organization, the client should first obtain the consent of HRC.

(iii) It also follows that original data which cannot be copied should not be removed from the control of HRC.

(d) Reports

The success of research resulting from testing facilities controlled by sponsors is no doubt indicated by the sales made by the company. The success of work performed by contract research organizations is, however, dependent on:

(i) the opinions of sponsors—who may develop personal relationships (trust) which will ensure future work, and

(ii) the report—which is the end-product and which persists and may be seen by governmental authorities, other 'sponsors', etc. To be successful, therefore, reports must be of high quality.

The presentation of reports is, therefore, of great importance, together with accuracy and professional interpretation. As regards accuracy, I believe that the introduction of QA units may be advantageous since they will add to the checking procedures of our reports, which will now consist of:

(i) spelling, punctuation and accuracy of typing . . . Proof readers

(ii) raw data transcription, by sampling* . . . QA unit

---

* In our opinion, it would be practically impossible to ensure that every fact and every figure recorded in every report had been doubly checked. Partial or complete computerization may reduce the size of the task but until this happens, some sampling technique for checking will be necessary.

(iii) intra-report transcription ... QA unit

(iv) scientific interpretation ... Principal Scientists

(v) over-view (at management level) ... Quality assurance

It has been proposed to us that it is the reports that should be sampled. It is our opinion, however, that all reports must be examined (by the QA unit) and the amount of transcribed original data, etc. that has to be checked, sampled.

COMPLIANCE WITH GLP

Perhaps the commonest question put to contract research organizations nowadays is: 'Can you give a written undertaking that this study will comply with the GLP?' Unfortunately, there is no simple answer and the alternatives are: 'yes', 'no' and 'maybe'.

(i) 'maybe' because, in fact there are no regulations at the present time and even later it may be difficult until standards for comparison have materialized—either by modification of the GLP or precedent.

(ii) 'no' because there is a natural tendency to set too high a target.

(iii) 'yes' because in the vast majority of testing facilities the standard of operation and the scientific level of competence is, I believe, still high.

MINOR ANOMALIES IN GLP

Consideration of the GLP in detail leads obviously to a number of anomalies which, compared with the principal danger, are relatively minor. A few examples have been included for the sake of completeness:

(i) How does one practically, give a *unique number* to small rodents, that will be useful over perhaps the lifespan of the animal?

(ii) What *illnesses* have to be reported to management to ensure that the sufferers in no way affect the animals on study?

(iii) At a practical level, what is considered as *adequate separation* of animals of the same species, when housed in the same room? (Already suggestions have been made that animals on one study, however small in number, should be housed in a separate room—whatever the expense. Is this scientifically necessary? Does not the control group, included in every study, act as an indicator of the 'normal state' for that species of animal?)

  (iv) As regards original data, it is suggested that *'magnetic tapes'* used for the purposes of dictation, may have to be preserved in the future. I noticed no reference to the shorthand notebooks of secretaries. The use of dictating machines is to increase efficiency and I can see little purpose in preserving tapes if the typed document is read, signed and dated by its author. (Incidentally, how many of us are so perfect when talking into a dictation machine, that we would want all comments preserved?)

  (v) An acclimatization period and evaluation, including veterinary examination, are required of all newly received animals: how can this be applied to the use of timed-pregnant rats?

  (vi) As regards feed and water, what is meant by, or how does one determine: known interfering contaminants, in order to demonstrate that they are not present?

  (vii) Finally, there have been many discussions on the subject of GLP requirements regarding test compounds, *mixtures* and the preservation of samples. To date, we have had no difficulty as all major clients have provided details of those compounds not involved in blind studies and they accept the responsibility for analysing mixture samples, since only they have experience with that particular test compound. A difficulty will, however, arise when a sponsor has no spare capacity in its own laboratories or a method for analysing the compound in, for example, diet, is not known. The expense of dealing with this situation, on some occasions, may be very high.

Before concluding, I would like to refer once again to the principal danger that I see associated with the present GLP Proposals. An analogy presented to me I thought particularly apt. Legally, in the UK, one may drive a car on dual carriageways at 60 miles an hour, but if the road is designated a 'motorway' the speed limit may be 70 miles an hour. The police are not supposed to stop you for speeding unless you have exceeded the stated limits. Consider now the position if the law merely stated: 'When driving a motor car on the highways in the United Kingdom, you may not drive *too fast*'.

10

Some problems in good laboratory practice for a contract research laboratory
Harald Reinert

I intend to comment on the impact of the proposed Good Laboratory Practice (GLP regulations) on the safety evaluation of drugs and to mention some laboratory practices which, in my view, are poor; to raise the problem of disqualification and suspension of a testing facility by a sponsor and to point to the inflationary effect of GLP regulations on the cost of toxicity studies, especially for contract research laboratories.

The GLP regulations have been drafted because studies have been found to be inadequately designed, conducted and reported. They are intended to eliminate the possibility that drugs and other chemicals are introduced when the quality of the non-clinical laboratory studies was poor or suspect and when reasonable care had not been taken to determine their toxic and carcinogenic risk or when safety issues are unresolved.

The GLP regulations will have little impact on the safety of drugs under chronic use conditions. This is because most of the clinically important toxic effects of drugs are and can only be discovered when they have been extensively used in man. The most obvious toxic, carcinogenic and teratogenic potential of new drugs will be uncovered in non-clinical tests and potentially dangerous compounds will be eliminated, or, should they be life-saving, then the knowledge of target organ toxicity will guide the clinician in the use of the drug. Those lesser toxic actions which become apparent only in man and only on long term use will at present not be discovered in non-clinical toxicity tests. Religious adherence to the GLP regulations will not answer the question: 'are adverse effects in man likely?'

Hardly any of the outstanding therapeutic advances, with the possible exception of beta-adrenoceptor blocking drugs and Histamine H_2 receptor antagonists, have been discovered by animal pharmacologists. For the most part it has been the astute reaction of clinicians to secondary or side effects which has marked the beginning of new, more important and more extensive therapeutic applications than those indications originally envisaged.

The same principle applies to the discovery in man of possible connections between chronic drug administration and adverse effects; for example such as breast tumour development and block of hypothalamic dopamine receptors, practolol and immune disorders, alpha-methyl-dopa and haemolysis, non-steroidal anti-inflammatory drugs and liver toxicity, oral contraceptives and myocardial infarction, hydralazine and rheumatic symptoms, clofibrate and thrombo-embolism, thiazides and gout, and many others.

None of these adverse effects can readily be observed in chronic animal studies and nothing, including the adherence to GLP regulations, will change this state of affairs until fundamental changes and scientific advances in drug safety evaluation have been made. The stereotype 6, 12, 18 or 24 months, four dose-level 2–3 species, safety tests will not find the toxic effects for whose discovery more than the usual routine procedures are required however excellent their quality.

There are some practices and experimental conditions in non-clinical safety evaluation which are debatable and require attention. For example some do not regard life span exposure necessary for the assessment of the carcinogenic risk of drugs. The use of small groups of rodents and exposure for periods substantially less than the lifespan of animals can not be good practice when there is evidence that with this procedure the carcinogenic potential of compounds can be missed. For example in one experiment of 18 months duration on 25 rats per sex and dose level there was nothing to suggest a carcinogenic potential and 3 benign exocrine pancreas tumours, were observed, one of which was in the control and the other one in the low dose group at 12 months and one in the same group at 18 months. The incidence was well within control experience. In a repeat experiment on 50 rats per sex and dose level and of 24 months duration a clearcut dose dependent increase in malignant and benign exocrine pancreas tumours, with none in the controls, was seen, with only 6 of the 58 tumours found before 24 months.

It is not generally appreciated that the composition of the diet can have dramatic effects on the pharmacology of a compound. For example Herken and co-workers (Herken, H., 1951) found that, in rats on a fat-free diet, the weak anticonvulsant activity of gamma-BHC was abolished. We know now that in linseed oil-treated rats cyclic-GMP and Prostaglandin F2α are decreased with a concomitant resistance to convulsants. The reverse may apply to a fat free diet so that rats on such a diet are more susceptible to convulsants and less ready subjects for the demonstration of anticonvulsant activity. Tannenbaum and Silverstone were the first to draw attention to the effect of diet composition on tumour development when they found that an increase in the protein content increased the incidence of hepatomas from 10 to 50% and a decrease of the protein concentration abolished hepatomas in males (Tannenbaum, A. and Silverstone, H., 1953).

The varying composition and possible contamination of diets with herbicides, pesticides, plant steroids or mycotoxins, can affect the frequency and type of spontaneous and induced tumours. For this reason the number of control animals should be double that of test groups and a continuous laboratory tumour control should be run and the diet routinely analysed for composition and major contaminants. Bedding would hardly be regarded as critical. However, the use of cedar shavings is known to increase, in mice, the mammary tumour incidence by a factor of ten.

Ammonia levels will rise in the now conventional shoe-box type plastic cages if the bedding material has poor absorbent properties. The ammonia levels could be further enhanced by the filter with which cages should be covered to avoid air dispersal and cross-contamination by the compound under test. Increased ammonia levels are known to decrease the activity of some microsomal enzymes and the phagocytic activity of macrophages.

I question whether urinalysis is a reasonable laboratory practice during chronic trials. Urine is collected from laboratory animals in so called metabolism cages. The end product consists of water, hair, faeces, epithelium, smegma, pus, vaginal discharge, sperm, seminal fluid, bacteria and urine. A quantitative analysis including osmolarity and microscopy is done on this 'fluid' and the reports filled with pages of results of more than doubtful value. Any renal impairment will be diagnosed with much better precision during histopathological examination and in conjunction with biochemical (clinical chemical) assays. Proteinuria in rats is a known pathophysiological phenomenon increasing with the age of the animals and an initial male sex specificity. This parameter is therefore of no value for the determination of renal impairment through drug action. We have shown huge fluctuation to occur from day to day in the same animal and between animals under strictly controlled environmental conditions. Whatever the causes, our studies showed that single 16-hour urine samples are not representative of renal function. Likewise, means established from a single animal over several consecutive days are of doubtful significance due to the large day-to-day variations. We have also shown that this is true for 24-hour samples. In addition the quantitative tests usually done (blood, sugar, proteins, crystals . . .) could be falsely negative in case of dilution by drinking water for example. I therefore assert that urinalysis in chronic toxicity studies is a bad routine practice. If measurement of renal function is required then this should be studied in a separate experiment with concentration-dilution renal clearance and other renal function tests and urine collected properly by catheterization.

In some study protocols it is required to weigh some 14 organs of rats and mice. These organs are meticulously trimmed and weighed and dry out in the process. This is detrimental to the quality of the histological specimens. Nothing is observed from organ weights which cannot be seen far more precisely under the microscope. An excellent quality of slide production is a better and far more important laboratory practice than the accumulation of results of organ weights which occupy the statistician but have in my experience never had a decision value in a drug safety issue.

Sophisticated random distribution is practiced for cage positioning. There is no evidence that a toxic or carcinogenic compound has been missed because random distribution of cages in space has not been observed.

After this scientific effort animals are bled by cutting the tip of the tail, sometimes even after previous heating of the rat to 40 or 50°C in a hot box to produce vasodilatation and an increased blood flow. This blood sample contains at least six ingredients other than blood; epithelium, lymph various cellular structures of bone, cartilage, tendon and muscle. No wonder that samples are often haemolysed. This is then compensated for either by 'grading' or by excessively large ranges of 'normal' values. There are many more examples of bad laboratory practices: the measurement of blood coagulation in untreated glass or plastic tubes, storage of specimens without knowledge of half-life or decay times, antiquated methodology—specific gravity, or refractometer for proteins, no attempts to determine the possibility of an interaction of drugs in plasma with parameters assayed in clinical pathology. The problem of conventional but debatable procedures causes me concern in the context of the GLP regulations sub-part on the disqualification of a testing facility, because a divergence of opinions between scientists and laboratories on what is a good and what is a bad method or laboratory practice can give a sponsor the right to suspend a non-clinical testing facility. I have no difficulties in seeing this part of the GLP regulations as wide open to misuse. The possibility exists in the GLP regulations to disqualify a testing facility because, in the view of a sponsor, a study has or is being carried out with an inadequate number or insufficiently trained personnel.

There will always be variations between different laboratories in the number of staff employed for a particular activity. This depends on the individual work tempo, attitudes, motivation, national characteristics, systems, methods used and the quality of the organization. The less well organized and equipped department requires more staff for the same volume of work and by comparison the better organized unit could be regarded as inadequately staffed. It would also be simple to hire key staff, especially those with scarcity value and difficult to replace, away from a testing facility and declare any study void because the number of personnel became inadequate.

Paragraph 3 e 202(d) specifies that the testing facility must conduct studies in physical facilities of suitable size and construction or location and design and that a facility shall be determined not suitable unless it provides these. The most disturbing aspect of this sub-part of the regulation is that the sponsor may at any time terminate or suspend a testing facility and need not utilize either the grounds or the procedures for disqualification set forth in the GLP regulations. I believe that any testing facility could be declared unsuitable by an unsympathetic sponsor or inspector and thus be suspended or disqualified because of what might be regarded as unsuitable, inadequate or inappropriate. I wonder whether even governmental laboratories might not fail in the face of hostile inspection. This sub-part can be exploited to the detriment of a testing facility and has to be re-written with explicit and concise statements and new unambiguous regulations.

It has been recognized by the authorities that the proposed regulations may have economic or inflationary implications for laboratories engaged in toxicity testing on animals. To start with there is added capital cost because no research establishment has substantial and spare storage areas available to accommodate, for periods of up to 10 years, all documents generated as well as biological specimens from toxicity, carcinogenicity, mutagenicity and teratogenicity testing. Some rebuilding and modification to existing structures has to be done and new areas have to be built for the additional staff required for adherence to the GLP regulations.

The procedures required by the GLP regulations in terms of checking, signing, recalculation, supervision, and confirmation add to the work load and operating cost. Microbiological assays and histopathological quality control has to be done on animal intake, quality assurance checks by the personnel in the toxicity, pathology and other departments, quality assurance inspections on experimental data, competence, dosing, bleeding, body weight recording, certification of records, certification and verification of all archival data, binding, logging, duplicating and so on. We estimate that some 140 hours have to be added to a 6-month-plus test or between 15–20% increase in the time spent on a study in the toxicology division alone. Added operational cost and overheads in terms of stationery, canteen facilities and other social services, cleaning, repair and maintenance will be in the region of 17%. We have estimated that we require an additional 14 people to deal with the same number of studies when applying the GLP regulations. In the Quality Assurance unit we need as a minimum a manager, an assistant and a secretary/clerk. For other units we require two graduates and four technicians for toxicity and carcinogenicity testing, two pathology and clinical pathology technicians, for the archives one graduate, one clerk/secretary and one storeman. This personnel increase will add 31% to the pay roll, in the toxicology group.

A great number of items will increase the cost of non-clinical testing. The extent of the increase will be largely unknown until experience has revealed the effective size of the cleaning, stores, clerical, technical, scientific, supervisory and quality assurance staff, the level of building and capital expenditure, stationery, social services, maintenance, repairs etc. There can be no doubt that the GLP regulations will have a pronounced inflationary impact on the cost of non-clinical laboratory studies. In conclusion I have four observations to make;

1. The most meticulous book keeping, the most thorough auditing and painstaking adherence to the GLP regulations will in no way influence the *scientific quality* of the methods, the observations or the observer.

The introduction of the GLP regulations will not overcome the limitations of animal experimentation in the evaluation of the potential hazards of new drugs. The procedures have been designed to answer several questions, to one of which: 'Are adverse effects likely?' The GLP regulations will, unfortunately, provide no answer. It is a historical fact that the clinically novel and important pharmacological effects of drugs have invariably been discovered in man after long term and extensive exposure.

2. The GLP regulations might improve the quality of experimentation provided reason and soundness has priority over quantity and insistance on adherence to traditional and conventional procedures some of which are of little value for the assessment of risk.

3. There is a genuine concern, especially in contract laboratories, about the misuse of the ambiguous and ill-defined aspects of the disqualification sub-part of the Regulations.

It is strongly urged that an entirely new sub-part in specific and precise terms be provided.

4. The GLP regulations have economic implications and the cost of non-clinical testing will increase because adherence to the regulations necessitates capital investment, recruitment of additional personnel and increased operational costs.

References

1. Herken, H., (1951). *Arzneim. Forsch.*, **1**, 356
2. Tannenbaum, A. and Silverstone, H., (1953). *Adv. Cancer Res.*, **1**, 451

Index